TEXTBOOK OF GREEN CHEMISTRY

TEXTBOOK OF GREEN CHEMISTRY

By

Dr. Shubhrata R. Mishra

Dept. of Botany
Vikram University
Ujjain (M.P.)
(India)

DISCOVERY PUBLISHING HOUSE PVT. LTD.
NEW DELHI-110 002

Published by:
Namit Wasan
DISCOVERY PUBLISHING HOUSE PVT. LTD.
4383/4B, Ansari Road, Darya Ganj
New Delhi-110 002 (India)
Phone : +91-11-23279245; 23253475; 43596065
E-mail : discoverybooksindia@gmail.com
discoverypublishinghouse@gmail.com
namitwasan9@gmail.com
web : www.discoverypublishinggroup.com

***First Edition:* 2011**
ISBN: 978-81-8356-840-1

Textbook of Green Chemistry

Preface

Green chemistry, also called 'sustainable chemistry', is a chemical philosophy encouraging the design of products and processes that reduce or eliminate the use and generation of hazardous substances. Whereas environmental chemistry is the chemistry of the natural environment, and of pollutant chemicals in nature, green chemistry seeks to reduce and prevent pollution at its source. In 1990 the Pollution Prevention Act was passed in the United States. This act helped create a *modus operandi* for dealing with pollution in an original and innovative way. It aims to avoid problems before they happen.

Green chemistry is that branch of chemistry,which deals with the study of design of products and processes that reduce or eliminate the use and generation of hazardous substances.

As a chemical philosophy, green chemistry derives from organic chemistry, inorganic chemistry, biochemistry, analytical chemistry, and even physical chemistry. However, the philosophy of green chemistry tends to focus on industrial applications. Click chemistry is often cited as a style of chemical synthesis that is consistent with the goals of green chemistry. The focus is on minimizing the hazard and maximizing the efficiency of any chemical choice. It is distinct from environmental chemistry which focuses on chemical phenomena in the environment.

In 2005 Ryoji Noyori identified three key developments in green chemistry: use of supercritical carbon dioxide as green solvent, aqueous hydrogen peroxide for clean oxidations and the use of hydrogen in asymmetric synthesis. Examples of applied green chemistry are supercritical water oxidation, on water reactions and dry media reactions.

Bioengineering is also seen as a promising technique for achieving green chemistry goals. A number of important process chemicals can be synthesized in engineered organisms, such as shikimate, a Tamiflu precursor which is fermented by Roche in bacteria.

Author

Contents

1

Introduction

Green chemistry, also called 'sustainable chemistry', is a chemical philosophy encouraging the design of products and processes that reduce or eliminate the use and generation of hazardous substances. Whereas environmental chemistry is the chemistry of the natural environment, and of pollutant chemicals in nature, green chemistry seeks to reduce and prevent pollution at its source. In 1990 the Pollution Prevention Act was passed in the United States. This act helped create a *modus operandi* for dealing with pollution in an original and innovative way. It aims to avoid problems before they happen.

Green chemistry is that branch of chemistry,which deals with the study of design of products and processes that reduce or eliminate the use and generation of hazardous substances.

As a chemical philosophy, green chemistry derives from organic chemistry, inorganic chemistry, biochemistry, analytical chemistry, and even physical chemistry. However, the philosophy of green chemistry tends to focus on industrial applications. Click chemistry is often cited as a style of chemical synthesis that is consistent with the goals of green chemistry. The focus is on minimizing the hazard and maximizing the efficiency of any chemical choice. It is distinct from environmental chemistry which focuses on chemical phenomena in the environment.

In 2005 Ryoji Noyori identified three key developments in green chemistry: use of supercritical carbon dioxide as green solvent, aqueous hydrogen peroxide for clean oxidations and the use of hydrogen in asymmetric synthesis. Examples of applied green chemistry are supercritical water oxidation, on water reactions and dry media reactions.

Bioengineering is also seen as a promising technique for achieving green chemistry goals. A number of important process chemicals can be synthesized in engineered organisms, such as shikimate, a Tamiflu precursor which is fermented by Roche in bacteria.

Principles

Paul Anastas, then of the United States Environmental Protection Agency, and John C. Warner developed 12 principles of green chemistry, which help to explain what the definition means in practice. The principles cover such concepts as:

- the design of processes to maximize the amount of raw material that ends up in the product;
- the use of safe, environment-benign substances, including solvents, whenever possible;
- the design of energy efficient processes;
- the best form of waste disposal: do not create it in the first place.

The 12 principles are:

- *Prevent waste:* Design chemical syntheses to prevent waste, leaving no waste to treat or clean up.

- *Design safer chemicals and products:* Design chemical products to be fully effective, yet have little or no toxicity.
- *Design less hazardous chemical syntheses:* Design syntheses to use and generate substances with little or no toxicity to humans and the environment.
- *Use renewable feedstock:* Use raw materials and feedstock that are renewable rather than depleting. Renewable feedstock are often made from agricultural products or are the wastes of other processes; depleting feedstock are made from fossil fuels (petroleum, natural gas, or coal) or are mined.
- *Use catalysts, not stoichiometric reagents:* Minimize waste by using catalytic reactions. Catalysts are used in small amounts and can carry out a single reaction many times. They are preferable to stoichiometric reagents, which are used in excess and work only once.
- *Avoid chemical derivatives:* Avoid using blocking or protecting groups or any temporary modifications if possible. Derivatives use additional reagents and generate waste.
- *Maximize atom economy:* Design syntheses so that the final product contains the maximum proportion of the starting materials. There should be few, if any, wasted atoms.
- *Use safer solvents and reaction conditions:* Avoid using solvents, separation agents, or other auxiliary chemicals. If these chemicals are necessary, use innocuous chemicals. If a solvent is necessary, water is a good medium as well as certain eco-friendly solvents that do not contribute to smog formation or destroy the ozone.
- *Increase energy efficiency:* Run chemical reactions at ambient temperature and pressure whenever possible.
- *Design chemicals and products to degrade after use:* Design chemical products to break down to innocuous

substances after use so that they do not accumulate in the environment.

- *Analyze in real time to prevent pollution:* Include in-process real-time monitoring and control during syntheses to minimize or eliminate the formation of byproducts.
- *Minimize the potential for accidents:* Design chemicals and their forms (solid, liquid, or gas) to minimize the potential for chemical accidents including explosions, fires, and releases to the environment.

The Presidential Green Chemistry Challenge Awards began in 1995 as an effort to recognize individuals and businesses for innovations in green chemistry. Typically five awards are given each year, one in each of five categories: Academic, Small Business, Greener Synthetic Pathways, Greener Reaction Conditions, and Designing Greener Chemicals. Nominations are accepted the prior year, and evaluated by an independent panel of chemists convened by the American Chemical Society. Through 2006, a total of 57 technologies have been recognized for the award, and over 1000 nominations have been submitted.

In 1996, Dow Chemical won the 1996 Greener Reaction Conditions award for their 100% carbon dioxide blowing agent for polystyrene foam production. Polystyrene foam is a common material used in packing and food transportation. Seven hundred million pounds are produced each year in the United States alone. Traditionally, CFC and other ozone-depleting chemicals were used in the production process of the foam sheets, presenting a serious environmental hazard. Flammable, explosive, and, in some cases toxic hydrocarbons have also been used as CFC replacements, but they present their own problems. Dow Chemical discovered that supercritical carbon dioxide works equally as well as a blowing agent, without the need for hazardous substances, allowing the polystyrene to be more easily recycled. The CO_2 used in the process is reused from other industries, so the net carbon released from the process is zero.

In 2002, Cargill Dow (now Natureworks) won the Greener Reaction Conditions Award for their improved polylactic acid polymerization process. Lactic acid is produced by fermenting corn and converted to lactide, the cyclic dimer ester of lactic acid using an efficient, tin-catalyzed cyclization. The L, L-lactide enantiomer is isolated by distillation and polymerized in the melt to make a crystallizable polymer, which has use in many applications including textiles and apparel, cutlery, and food packaging. Wal-Mart has announced that it is using/will use PLA for its produce packaging. The NatureWorks PLA process substitutes renewable materials for petroleum feedstocks, doesn't require the use of hazardous organic solvents typical in other PLA processes, and results in a high-quality polymer that is recyclable and compostable.

In 2003 Shaw Industries was recognized with the Designing Greener Chemicals Award for developing EcoWorx Carpet Tile. Historically, carpet tile backings have been manufactured using bitumen, polyvinyl chloride (PVC), or polyurethane (PU). While these backing systems have performed satisfactorily, there are several inherently negative attributes due to their feedstocks or their ability to be recycled. Shaw selected a combination of polyolefin resins as the base polymer of choice for EcoWorx due to the low toxicity of its feedstocks, superior adhesion properties, dimensional stability, and its ability to be recycled. The EcoWorx compound also had to be designed to be compatible with nylon carpet fiber. Although EcoWorx may be recovered from any fiber type, nylon-6 provides a significant advantage. Polyolefins are compatible with known nylon-6 depolymerization methods. PVC interferes with those processes. Nylon-6 chemistry is well-known and not addressed in first-generation production. From its inception, EcoWorx met all of the design criteria necessary to satisfy the needs of the marketplace from a performance, health, and environmental standpoint. Research indicated that separation of the fiber and backing through elutriation, grinding, and air separation proved to be the best way to recover the face and backing components, but an infrastructure for returning postconsumer EcoWorx to the elutriation process was necessary. Research also indicated that the postconsumer carpet tile had a positive economic value at the end of its useful life. EcoWorx is recognized by MBDC as a certified Cradle to Cradle design.

In 2005, Archer Daniels Midland (ADM) and Novozymes N.A. won the Greener Synthetic Pathways Award for their enzyme interesterification process. In response to the U.S. Food and Drug Administration (FDA) mandated labeling of *trans*-fats on nutritional information by January 1, 2006, Novozymes and ADM worked together to develop a clean, enzymatic process for the interesterification of oils and fats by interchanging saturated and unsaturated fatty acids. The result is commercially viable products without *trans*-fats. In addition to the human health benefits of eliminating *trans*-fats, the process has reduced the use of toxic chemicals and water, prevents vast amounts of byproducts, and reduces the amount of fats and oils wasted.

In 2006, Professor Galen J. Suppes, from the University of Missouri–Columbia, was awarded the Academic Award for his system of converting waste glycerin from biodiesel production to propylene glycol. Through the use of a copper-chromite catalyst, Professor Suppes was able to lower the required temperature of conversion while raising the efficiency of the distillation reaction. Propylene glycol produced in this way will be cheap enough to replace the more toxic ethylene glycol that is the primary ingredient in automobile antifreeze.

The Royal Australian Chemical Institute (RACI) presents Australia's Green Chemistry Challenge Awards. This awards program is similar to that at the United States Environmental Protection Agency? (EPA), although the Institute has included a category for Green Chemistry education as well as Small Business and Academic or Government.

The Canadian Green Chemistry Medal is an annual award given to an individual or group for promotion and development of green chemistry in Canada and internationally. The winner is presented with a citation recognizing the achievements together with a sculpture.

Green Chemistry activities in Italy center around an inter-university consortium known as INCA. Beginning in 1999, the INCA has given three awards annually to industry for applications of green chemistry. The winners receive a plaque at the annual INCA meeting.

In Japan, The Green and Sustainable Chemistry Network (GSCN), formed in 1999, is an organization consisting of representatives from chemical manufacturers and researchers. In 2001, the organization began an awards program. GSC Awards are to be granted to individuals, groups or companies who greatly contributed to green chemistry through their research, development and their industrialization. The achievements are awarded by Ministers of related government agencies.

In the United Kingdom, the Crystal Faraday Partnership, a non-profit group founded in 2001, awards businesses annually for incorporation of green chemistry. The Green Chemical Technology Awards have been given by Crystal Faraday since 2004; the awards were presented by the Royal Society of Chemistry prior to that time. The award is given only to a single researcher or business, while other notable entries are given recognition as well.

The Nobel Prize Committee recognized the importance of green chemistry in 2005 by awarding Yves Chauvin, Robert H. Grubbs, and Richard R. Schrock the Nobel Prize for Chemistry for "the development of the metathesis method in organic synthesis." The Nobel Prize Committee states, "this represents a great step forward for 'green chemistry', reducing potentially hazardous waste through smarter production. Metathesis is an example of how important basic science has been applied for the benefit of man, society and the environment."

Trends

Attempts are being made not only to quantify the *greenness* of a chemical process but also to factor in other variables such as chemical yield, the price of reaction components, safety in handling chemicals, hardware demands, energy profile and ease of product workup and purification. In one quantitative study, the reduction of nitrobenzene to aniline receives 64 points out of 100 marking it as an acceptable synthesis overall whereas a synthesis of an amide using HMDS is only described as adequate with a combined 32 points.

Green chemistry is increasingly seen as a powerful tool that researchers must use to evaluate the environmental impact of nanotechnology As nanomaterials are developed, the environmental and human health impacts of both the products themselves and the processes to make them must be considered to ensure their long-term economic viability.

Laws

In 2007, Europe put into place the Registration, Evaluation, Authorisation, and Restriction of Chemicals (REACH) program, which requires companies to provide data showing that their products are safe. The US Toxic Substances Control Act, passed in 1976, has no similar provisions.

On September 29, 2008 California approved two laws which encourage green chemistry, launching the California Green Chemistry Initiative. The law requires California's Department of Toxic Substances Control to prioritize "chemicals of concern", and puts the burden of testing on the agency rather than industry. The laws were criticized by Paul Anastas, who stated that the laws did not go far enough in encouraging research, education, and industry incentives.

Research is currently ongoing in the area of supramolecular chemistry to develop reactions which can proceed in the solid state without the use of solvents. The cycloaddition of trans-1, 2-bis(4-pyridyl)ethylene is directed by resorcinol in the solid state. This solid-state reaction proceeds in the presence of UV light in 100% yield.

Reducing Market Barriers

In March 2006, the University of California published a report by Dr. Michael P. Wilson and colleagues, Daniel A. Chia and Bryan C. Ehlers, on potential chemicals policy for the California Legislature entitled, Green Chemistry in California: A Framework for Leadership in Chemicals Policy and Innovation. The report finds that long-standing weaknesses in the U.S. chemical management program, notably the Toxic Substances Control Act (TSCA) of 1976, have produced a chemicals market in the U.S. that discounts the hazardous properties of chemicals relative to their function, price, and performance. The report concludes that these market conditions represent a key barrier to the scientific, technical, and commercial success of green chemistry in the U.S., and that fundamental policy changes are needed to correct these weaknesses.

The term *green chemistry,* coined in 1991, is defined as "the design of chemical products and processes that reduce or eliminate the use and generation of hazardous substances." This approach to the protection of human health and the environment represents a significant departure from the traditional methods previously used. Although historically societies have tried to minimize exposure to chemicals, green chemistry emphasizes the design and creation of chemicals that are not hazardous to people or the environment. It has been applied to a wide range of industrial and consumer goods, including paints, dyes, fertilizers, pesticides, plastics, medicines, electronics, dry cleaning, energy generation, and water purification.

At the heart of green chemistry is the recognition that hazard is simply another property of a chemical substance. Properties of chemicals are caused by their molecular structure; they can be modified by changing that structure. The types of hazards that can be addressed by green chemistry vary. They include physical hazards (being explosive or flammable), toxicity (being carcinogenic or cancer causing, or lethal), or global hazards (climate change or stratospheric ozone depletion). Therefore, in the same way that a substance can be designed to be red or hard, it may also be designed to be nontoxic.

Chemists and chemical engineers applying green chemistry look at the entire life cycle of a product or process, from the origins of the materials used for manufacturing to the ultimate fate of the materials after they have finished their useful life. By using such an approach, scientists have been able to reduce the impacts of harmful chemicals in the environment.

Research and development in the field of green chemistry are occurring in several different areas.

Alternative feedstocks

Historically, many of the materials used to make products often were toxic or depleted limited resources such as petroleum, but green chemistry research is developing ways to make products from renewable and nonhazardous substances, such as plants and agricultural wastes. For example, cellulose and hemicellulose, which constitute up to eighty percent of biomass, can be broken down to sugars, then fermented to chemical commodities such as ethanol, organic acids, glycols, and aldehydes. Converting biomass to ethanol has become economically and technically viable due to a new class of genetically modified bacteria capable of breaking down the different sugars in hemicellulose.

Benign Manufacturing

The methods used to make chemical materials, called synthetic methods, have often employed toxic chemicals such as cyanide or chlorine. In addition, these methods have at times generated large quantities of hazardous wastes. Green chemistry research is developing new ways to make these synthetic methods more efficient and to minimize wastes while also ensuring that the chemicals used and generated by these methods are as nonhazardous as possible. For example, a number of industries, such as the pulp and paper industry, use chlorine compounds in processes that generate toxic chlorinated organic waste. Green chemists have developed a new technology that converts wood pulp into paper using oxygen, water and polyoxometalate salts, while producing only water and carbon dioxide as by-products.

Designing Safer Chemicals

Once it is certain that the feedstocks and methods needed to make a substance are environmentally benign, it is important to ensure that the end product is as nontoxic as possible. By understanding what makes something harmful (the field of molecular toxicology), scientists are able to design the molecular structure so that it is not dangerous.

Green Analytical Chemistry

The detection, measurement, and monitoring of chemicals in the environment through analytical chemistry have long been a tool for environmental protection. Instead of measuring environmental problems after they occur, however, green chemistry seeks to prevent the formation of toxic substances and thus prevent such problems. By making sensors and other instruments part of industrial manufacturing processes, green analytical chemistry is able to detect even tiny amounts of a toxic substance and to adjust process controls to minimize or stop its formation altogether. In addition, although traditional methods of analytical chemistry employ substances such as hazardous solvents, green analytical methods are being developed to minimize the use and generation of these substances while conducting analysis.

Green chemistry is effective in reducing the impact of chemicals on human health and the environment. In addition, many companies have found that it can be cheaper and even profitable to meet environmental goals. Profits derive from higher efficiency, less waste, better product quality, and reduced liability. Many environmental laws and regulations target hazardous chemicals, and following all these requirements can be complicated. But green chemistry allows companies to comply with the law in much simpler and cheaper ways. Finally, green chemistry is a fundamental science-based approach. Addressing the problem of hazard at the molecular level, it can be applied to all kinds of environmental issues.

Since 1991, there have been many advances in green chemistry, in both academic research and industrial implementation. For example, Spinosad, an insecticide manufactured by fermenting a naturally occurring soil organism, was registered by the EPA as a reduced-risk insecticide in 1997. Spinosad does not leach, bioaccumulate, volatilize, or persist in the environment and in field tests left 70 to 90 percent of beneficial insects unharmed. It has a relatively low toxicity to mammals and birds and is slightly to moderately toxic to aquatic organisms, but is toxic to bees until it dries. In another advance, an industrial cleaning solvent, ethyl lactate, made from cornstarch and soybean oil was patented in 2000 and is competitively priced with petrochemical solvents. It biodegrades to carbon dioxide and water and has no known harmful effects for the environment, humans, or wildlife. These advances, however, represent an extremely small fraction of the potential applications of green chemistry. Because the products and processes that form the basis of the economy and infrastructure are based on the design and utilization of chemicals and materials, the challenges facing this field are enormous.

2

Green Chemistry in Practice

Green chemistry is both the chemistry of the future and the chemistry of today. It is based on a number of principles that ensure that both processes and end products are clean and safe. Green chemistry aims to conserve both energy and raw materials. In practice, this means that 'green' processes are often cheaper than conventional methods. Some current processes are already 'green', and the use of green chemistry is growing because it is environmentally friendly, and also because of legislation and international agreements that aim to reduce pollution. One of the basic ideas of green chemistry is to *prevent* production of hazardous and polluting materials rather than producing them and then cleaning up.

Green Chemistry:

- is safe;
- conserves raw materials and energy; and
- is more cost-effective than conventional methods.

Approaches to making chemical processes 'greener' include:

- redesigning production methods to use different starting materials;
- using different reaction conditions, catalysts, solvents etc; and
- using production methods with fewer steps.

Yield and Atom Economy

One of the key ideas of green chemistry is that of atom economy. This considers how much of the reactants in a chemical reaction end up in the final useful product or products. Ideally all the atoms of the reactants would end up incorporated into useful products – such a reaction would produce no waste at all – but this is rarely possible. The atom economy (also called atom utilisation) of a reaction is a measure of the percentage of the starting materials that actually ends up in the useful products.

The Yield of a Chemical Reaction

Before looking at atom economy, we will look at a similar idea that you will probably already be familiar with, that of yield.

Much of the chemical industry is concerned with turning one material (the raw material) into another that is more useful and that is worth more (the product). This is why the chemical industry is useful and is how it makes a profit – by turning low value materials into high value ones. This process often requires several steps and is called a chemical synthesis ('synthesis' means 'building up'). Naturally, the chemists who design these syntheses wish to make the maximum amount of product that they can from a given amount of raw material. Unfortunately, few chemical reactions go 100% to completion. Firstly, the reaction may be reversible and form an equilibrium mixture, secondly there may be two or more alternative products that can be produced from the same starting materials. Furthermore, there will inevitably be losses during the process of separating the desired product from other waste materials (called by-products).

Example 1: Formation of an equilibrium mixture – making the ester ethyl ethanoate ($CH_3COOC_2H_5$) from ethanol (C_2H_5OH) and ethanoic acid (CH_3COOH).

The following equilibrium is set up:

$$C_2H_5OH(l) + CH_3COOH \rightleftharpoons CH_3COOC_2H_5(aq) + H_2O(l)$$

At room temperature, if we mix 1 mol of ethanol with 1 mol of ethanoic acid, we eventually produce a mixture containing 0.33 mol of ethyl ethanoate, 0.33 mol of water, 0.67 mol of ethanol and 0.67 mol of ethanoic acid. This means that the reaction has not gone to completion. If it had gone to completion, we would have formed 1 mol of ethyl ethanoate and 1 mol of water and there would have been no ethanol and ethanoic acid left.

Example 2: Formation of alternative products - the reaction of 2-bromo-2-methylpropane [$(CH_3)_2CBrCH_3$) with sodium hydroxide

Here two reactions are possible, forming 2-methylpropan-2-ol [$(CH_3)_2COHCH_3$] and 2-methylpropene [$(CH_3)_2C{=}CH_2$) respectively:

1. $(CH_3)_2CBrCH_3 + NaOH = (CH_3)_2COHCH_3 + NaBr$
2. $(CH_3)_2CBrCH_3 + NaOH = (CH_3)_2\,C = CH_2 + NaBr + H_2O$

Under certain conditions, 1 mol of 2-bromo-2-methylpropane produces 0.9 mol of methylpropene and 0.1 mol of 2-methylpropan-2-ol.

In these sorts of situations, chemists talk about the yield of a chemical reaction. This is the number of moles obtained of a specified product divided by the maximum number of moles that could have been obtained if all the reactant had been converted to product. It is usually expressed as a percentage. So

$$\text{Yield} = \frac{\text{actual no. of moles of product}}{\text{maximum no. of moles of product}} \times 100\%$$

So in example 1

$$\text{Yield of ethyl ethanoate} = \frac{0.33 \text{ mol}}{1 \text{ mol}} \times 100\% = 33\%$$

In example 2

$$\text{Yield of methylpropene} = \frac{0.9 \text{ mol}}{1 \text{ mol}} \times 100\% = 90\%$$

There is a complication with the idea of yield. In both the above examples we mixed the reactants in exactly the right ratios to react as shown by the equation - one mole of each reactant in each case. If we have more of one reactant than is needed then we must work out the yield with respect to the reactant that is in short supply (we call this the limiting reactant) because it is this reactant that governs the maximum amount of product we can make.

In the second example, we could have mixed 1 mol of 2-bromo-2-methylpropane with 2, 3, or even 10 mol of sodium hydroxide and the maximum possible amount of methylpropene would still have been 1 mol. But if we had used 0.5 mol of sodium hydroxide, and 1 mol of 2-bromo-2-methylpropane, the sodium hydroxide would have become the limiting reactant and the maximum possible amount of methylpropene would have become 0.5 mol.

In practice, no chemical reaction ever produces a 100% yield. Even if the reaction goes to completion and there is only a single possible product, there are always losses involved in the practical processes of separating the required product from the rest of the reaction mixture. For example when filtering there will always be some material left on the filter paper, when pouring there will always be droplets left behind in the original container and so on. In the laboratory much of the skill of the practical chemist is in increasing the yield of a reaction by careful manipulation of the apparatus so as to minimise losses such as these. Industrially, good design of the process will minimise losses of this type.

Consider the reaction of ethanoic acid with sodium hydroxide: $CH_3COOH + NaOH = CH_3COONa + H_2O$

(a) How many moles of sodium hydroxide could react with one mole of ethanoic acid?

(b) Work out the relative molecular masses of:

  (i) ethanoic acid.

  (ii) sodium hydroxide.

(c) If 2 mol of ethanoic acid are mixed with 1 mol of sodium hydroxide, what is the limiting reactant?

(d) If 3.0 g of ethanoic acid are mixed with 4.0 g of sodium hydroxide, what is the limiting reactant?

(e) In the reaction described in (d), what is the maximum mass of sodium ethanoate that could be produced?

(f) In the reaction described in (d), 3.6 g of sodium ethanoate was separated from the reaction mixture. What is the yield?

Consider the reaction of sulfuric acid with sodium hydroxide: $H_2SO_4 + 2NaOH = Na_2SO_4 + 2H_2O$

(a) How many moles of sodium hydroxide could react with one mole of sulfuric acid?

(b) Work out the relative molecular masses of:

  (i) sulfuric acid.

  (ii) sodium hydroxide.

(c) If 2 mol of sulfuric acid are mixed with 2 mol of sodium hydroxide, what is the limiting reactant?

(d) If 6.0 g of sulfuric acid are mixed with 4.0 g of sodium hydroxide, what is the limiting reactant?

(e) In the reaction described in (d), what is the maximum mass of sodium sulfate that could be produced?

(f) In the reaction described in (d), 5.0 g of sodium sulfate was separated from the reaction mixture. What is the yield?

Consider the oxidation of ethanol (C_2H_5OH) to ethanal (CH_3CHO) by acidified potassium dichromate(VI) ($K_2Cr_2O_7$) solution.

(a) Write an equation for the reaction. Use [O] to represent the oxidising agent potassium dichromate(VI).

(b) What other organic product is possible?

(c) Explain why this reaction will not have a 100% yield.

The idea of yield is a useful one but, from the point of view of green chemistry, it does not tell the whole story. This is because it concentrates only on one reactant (the limiting reactant) and one product (the product that we want). One of the key principles of green chemistry is that processes should be designed to incorporate the maximum amount of *all* the raw materials into the final product, thus reducing waste products. A reaction with a high percentage yield does not necessarily fulfil this principle.

Take as an example the reaction between ethyl propanoate ($CH_3CH_2COOCH_2CH_3$) and methylamine (CH_3NH_2) to form N-methylpropanamide ($CH_3CH_2CONHCH_3$) and ethanol (C_2H_5OH). 1 mol of each reactant and product is involved. The mass of 1 mol of each reactant (*ie* the relative molecular mass in grams) is given, but you might like to check these in a data book or by adding up the relative atomic masses for each compound.

$CH_3CH_2COOCH_2CH_3 + CH_3NH_2 = CH_3CH_2CONHCH_3 + C_2H_5OH$

1 mol	1 mol	1 mol	1 mol
118 g	31 g	103 g	46 g

Imagine that the reaction has a 100% yield and then look at the quantities involved. Assuming that N-methylpropanamide is the product we want, we have started with 149 g of starting materials (118 g + 31 g) yet the mass of the required product is only 103 g. The rest, 46 g of ethanol, is wasted. Looking at it another way, we have lost 2 atoms of carbon, 6 of hydrogen and one of oxygen. These are wasted and have to be disposed of. This, of course, is before we have taken into account any losses associated with the reaction not going to completion, alternative products and practical manipulation.

If the reaction were being used industrially, and it was economically viable to collect and purify the ethanol for use, then the picture would change; the atom economy could then be considered to be 100%.

It may be easier to see what has happened if we look at the displayed formulae of the molecules involved and colour code the atoms. Those coloured in green are incorporated into the final product and those in red are wasted. We could imagine a 'waste box' containing two atoms of carbon, six atoms of hydrogen and one atom of oxygen.

Green chemists have developed the idea of atom economy or atom utilization. This is defined as

% Atom economy =

$$\frac{\text{mass of desired product}}{\text{mass of desired product plus waste products}} \times 100\%$$

$$\text{\% Atom economy} = \frac{\text{mass of desired product}}{\text{total mass of reactants}} \times 100\%$$

which amounts to exactly the same thing, see below:

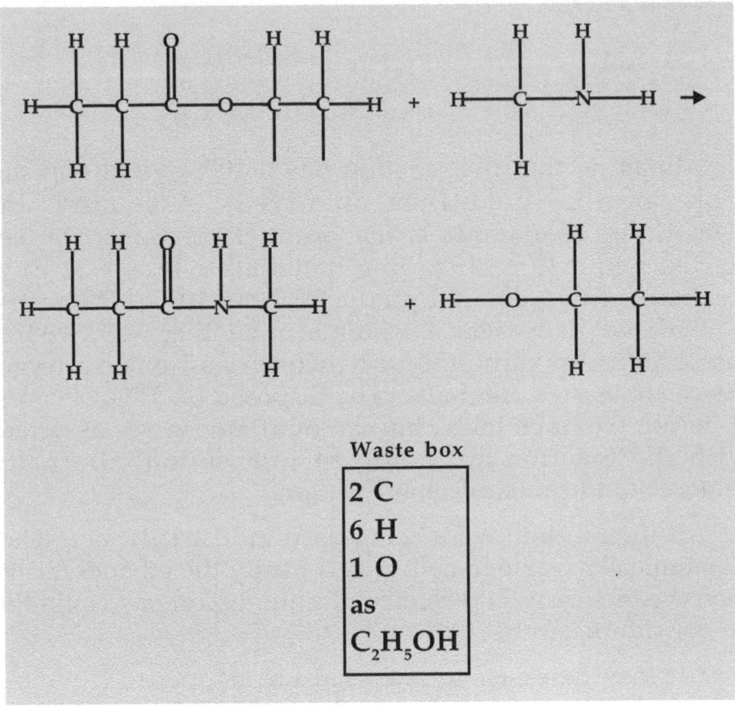

Fig. 2.1

Explain why 'mass of desired product plus waste products' and 'total mass of reactants' have exactly the same value.

In the example above of the reaction between ethyl propanoate and methylamine,

$$\text{the atom economy} = \frac{103}{149} \times 100\% = 69.1\%$$

So 69.1% of the starting materials are incorporated into the desired product, the rest is lost or, worse, has to be disposed of.

Now consider a similar reaction but starting with *methyl* propanoate ($CH_3CH_2COOCH_3$) rather than ethyl propanoate. This gives the same product, N-methylpropanamide, but with methanol as a by-product.

$CH_3CH_2COOCH_3$ + CH_3NH_2 = $CH_3CH_2CONHCH_3$ + H_3OH

1 mol	1 mol	1 mol	1 mol
104 g	31 g	103 g	32 g

$$= \frac{103}{135} \times 100 = 76.3\%$$

Atom economy

In this case, the 'waste box' contains one atom of carbon, four atoms of hydrogen and one atom of oxygen.

So a small change in the design of the synthesis (just shortening the carbon chain of one of the starting materials in this case) can lead to a significant improvement in the atom economy. In this case a molecule of methanol (mass 32 g mol^{-1}) is the unwanted product rather than a molecule of ethanol (mass 46 g mol^{-1}).

(a) Write out the equation for the reaction of propanoic acid (CH_3CH_2COOH) and methylamine to form N-methylpropanamide ($CH_3CH_2CONHCH_3$).

(b) (i) What molecule is the unwanted product?

(ii) Therefore what atoms are in the 'waste box'?

(c) Without doing a calculation, predict how the atom economy will compare with the above reaction.

(d) Now check your answer by calculating the atom economy.

Reaction Type and Atom Economy

Chemical reactions are often classified as addition, elimination, substitution or re-arrangement. Some examples of each of these reaction types follow, with their atom economies calculated. In each case, the mass of 1 mol (the relative molecular mass in grams) of each species is given below its formula.

Addition

$$CH_2 = CH_2 + Br_2 = CH_2BrCH_2Be$$

28	160	188

$$\text{Atom economy} = \frac{188}{(28+160)} \times 100\% = 100\%$$

You should be able to see without doing a calculation that the atom economy of *all* addition reactions is 100% because there is only one product and nothing is wasted.

Elimination

$$C_2H_5OH = CH_2 = CH_2 + H_2O$$

46	28	18

$$\text{Atom economy} = \frac{28}{46} \times 100 = 60.9\%$$

The atom economy of an elimination reaction can *never* be 100% because there is *always* a molecule wasted.

Substitution

$$CH_3COCl + C_2H_5OH = CH_3COOC_2H_5 + HCl$$

78.5	46	88	36.5

$$\text{Atom economy} = \frac{88}{(78.5 + 46)} \times 100\% = 70.7\%$$

The atom economy of a substitution reaction must always be less than 100% because there is always a molecule wasted. The atom economy is relatively low in this case because of the heavy chlorine atom that is ejected as part of the HCl.

(a) The main product of the reaction above is ethyl ethanoate ($CH_3COOC_2H_5$) Write the equation for the reaction of ethanoic acid with ethanol that also produces ethyl ethanoate as the main product.

(b) Work out the atom economy of the reaction you have suggested, to confirm that it is greater than that of the reaction above that starts with ethanoyl chloride (CH_3COCl).

(c) Explain why this reaction has a greater atom economy.

(d) What disadvantages does the reaction starting from ethanoic acid have if you wish to prepare ethyl ethanoate?

Rearrangement

$$CH_3CH_2CH_2OH = CH_3CHOHCH_3$$

60	60

$$\text{Atom economy} = \frac{60}{60} \times 100\% = 100\%$$

The atom economy of all re-arrangement reactions is 100% as nothing is lost.

The above rearrangement reaction cannot easily be carried out in one step but it can be done in two.

(a) Suggest a reaction scheme for bringing it about in two steps.

(b) What is the probable effect on the yield of having to use two steps rather than one? Explain your answer.

(a) For each of the reactions below, draw the displayed formula (a formula showing every atom and every bond) of each reactant and product. Colour each atom that appears in the final product in green and each atom that does not in red. For each reaction, say what atoms would be in the 'waste box'. (Check also that you can name each of the organic compounds.)

(i) $C_2H_5OH = CH_3OCH_3$

(ii) $CH_3COCH_3 + I_2 = CH_3COCH_2I + HI$

(iii) $C_6H_6 + 3H_2 = C_6H_{12}$

(iv) $C_2H_5I = CH_2 = CH_2 + HI$

(b) Work out the atom economy for each of the reactions above. In each case assume that the first mentioned product (if there is more than one) is the one that is required.

(c) Classify each reaction as one of the following types - addition, elimination, substitution and re-arrangement

(d) (i) For reaction 10 (a) (iv), suggest a different starting material that would lead to the same organic product but with better atom economy.

(ii) Work out the atom economy of your proposed reaction to confirm that its value is lower than that for the reaction in 10 (a) (iv).

The key reaction in the Haber process for the manufacture of ammonia is $N_2 + 3H_2 \rightleftharpoons 2NH_3$

(a) Work out the atom economy for this reaction. (You should not need to do a calculation.)

(b) What does the symbol $\rightleftharpoons$ suggest about the likely yield for the reaction?

Selectivity

Another concept, related to yield and atom economy that is used in synthetic organic chemistry is that of selectivity. Selectivity is concerned with controlling reactions so that the desired product is produced rather than competing side products.

The reaction of 2-bromo-2-methylpropane [$(CH_3)_2CBrCH_3$] with sodium hydroxide discussed above illustrates this idea. As we have seen, there are two possible products: 2-methylpropan-2-ol [$(CH_3)_2COHCH_3$] and methylpropene [$(CH_3)_2C = CH_2$]. By changing the conditions of the reaction, it is possible to control the proportions of each that are produced. Carrying out the reaction at a relatively low temperature in aqueous solution favours the formation of the alcohol by reaction (2.1) while at high temperature and in alcoholic solution the formation of the alkene predominates by reaction (2.2).

$$(CH_3)_2CBrCH_3 + NaOH = (CH_3)_2COHCH_3 + NaBr \quad (2.1)$$

$$(CH_3)_2CBrCH_3 + NaOH = (CH_3)_2\,C = CH_2 + NaBr + H_2O \quad (2.2)$$

Classify each of the above reactions as addition, elimination, substitution or re-arrangement.

With the conditions under which 90% of methylpropene and 10% of 2-methylpropan-2-ol are produced, this is a fairly selective reaction. If 50% of each product were formed, it would be entirely unselective. Clearly, chemists favour selective reactions when designing syntheses.

Chemists distinguish different types of selectivity.

Chemoselectivity is the ability of a reagent to react with one functional group rather than another. For example sodium tetrahydridoborate(III) (sodium borohydride, $NaBH_4$) is a selective reducing agent; it will reduce C=O but not C=C. Hydrogen with a platinum or nickel catalyst is less selective; it will reduce both these functional groups.

(a) Predict the products when CH_2=$CHCH_2CHO$ is reduced with:

(i) sodium tetrahydridoborate(III); and

(ii) hydrogen with a nickel catalyst.

(b) Sodium tetrahydridoborate(III) generates the ion H- and it is this that carries out the first step of the reduction.

(i) Which of the functional groups C=O or C=C is polar?

(ii) Mark the polarity of this functional group using the symbols d^+ and d^-.

(iii) Use this to explain why sodium tetrahydridoborate(III) will reduce C=O but not C=C.

(iv) On the basis of your explanation predict whether sodium tetrahydridoborate(III) will reduce the -CN functional group.

Regioselectivity is the ability to produce one positional isomer rather than another. (Positional isomers differ in the point on the carbon chain where a functional group is attached.) The addition of reagents H-X across a double bond is an example of a regioselective reaction. Markovnikov's rule states that the hydrogen will add to the carbon that already has more hydrogen atoms bonded to it.

(a) What are the two possible products when hydrogen bromide adds on to propene (CH_3CH=CH_2)?

(b) Which one does Markovnikov's rule predict will predominate? Hint: it will almost certainly help if you first draw out the displayed formulae (formulae showing every atom and every bond) of the two possible products.

Enantioselectivity is the ability of a reaction to produce one optical isomer rather than another. Optical isomers are pairs of mirror image isomers – they have virtually identical physical and chemical properties but often differ in their biological effects. (They can be distinguished by the fact that they rotate the plane of polarisation of polarised light in different directions.)

Most chemical reactions that make optical isomers produce a mixture containing equal amounts of the two isomers (called a racemic mixture), but reagents have been developed that can produce one rather than another. This is important in pharmaceutical chemistry because there are many cases where one of a pair of optical isomers is a safe and useful drug and the other has no activity or is even toxic. Ibuprofen (sold under the brand names Nurofen, Cuprofen and others) is a good example; one of the two enantiomers (called the *S*-form) is an effective pain-killing drug, the other (the *R*-form) is inactive. Because of the difficulty of separating the two, the drug is normally supplied as racemic (50 : 50) mixture of the two forms – an obvious source of waste which could be avoided if an enantioselective synthesis were used. There is more information about the synthesis of ibuprofen in *Ibuprofen – a case study in green chemistry*.

There are even more extreme examples where one optical isomer is a useful drug and the other is toxic. An example of this is penicillamine (not to be confused with penicillin). One optical isomer is an anti-arthritis drug, while its mirror image is toxic. Penicillamine is currently available as a racemic mixture but must be used with caution because one isomer is toxic to the heart and nervous system.

Enantioselective reactions are important for green chemistry because even if a racemic mixture can be separated, the unwanted component normally has to be thrown away.

Recently an enantioselective synthesis of the anti-inflammatory drug Naproxen has been developed using a special catalyst to add hydrogen across the double bond of a starting material in such a way that only one of the two possible optical isomers is produced, see Fig. 2.2.

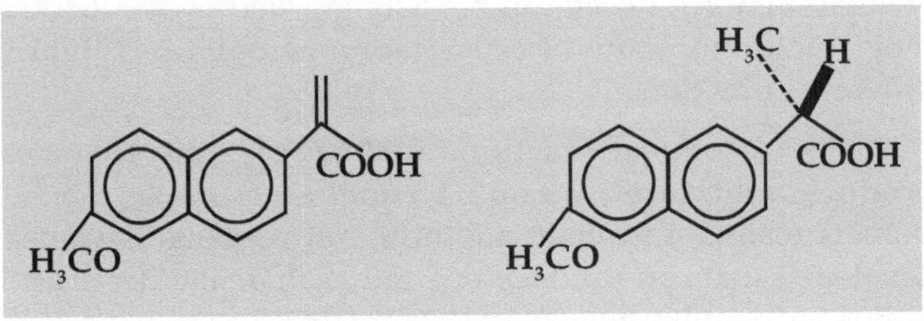

Fig. 2.2

Part of the synthesis of Naproxen

However, at present, Naproxen is still made by a non-selective route and sold as a racemic mixture because changing the synthetic route would require expensive and time-consuming toxicological tests to be repeated.

Other chiral drugs, such as the penicillins, are made and sold as single optical isomers but they are made from naturally-occurring starting materials that are already single optical isomers.

If an organic molecule has a carbon atom to which four different atoms or groups are bonded, it will exist as a pair of optical isomers. Such a carbon atom is called a chiral centre.

The skeletal formula of ibuprofen is given below.

(a) Draw the displayed formula (a formula showing every atom and every bond) of ibuprofen.

(b) Identify the chiral centre and mark it with a * on the formula.

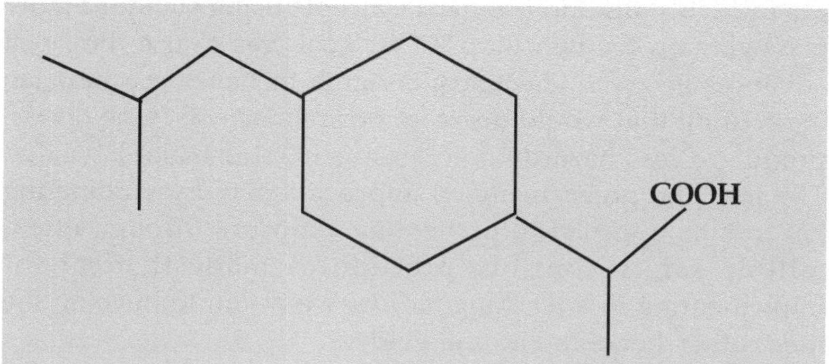

Fig. 2.3

Many of us are have struggled with the problem of how to modify our organic laboratory curriculum to reduce the amount of waste generated and make the lab safer for our students. At the University of Oregon we have developed a new organic chemistry lab curriculum based upon the principles of "green" chemistry1 to address these challenges. describe both the practical advantages of this curriculum and suggest that it provides outstanding opportunities to positively influence students' opinions about chemistry and encourage them to take an active part in shaping the future of chemistry in our society.

The traditional approaches to minimizing waste and improving safety in the organic lab involve microscale apparatus and increased ventilation. We initially pursued this approach, building a new laboratory where each student could perform his/her work with microscale glassware inside a fume hood. Although these changes reduced the amount of waste and provided a safer working environment, we were frustrated by two major disadvantages. A practical problem was that the lab's 18-student capacity meant night and weekend labs for some students. A more fundamental problem was that our reliance on microscale methods was not adequately preparing our students for working in our research labs or in industry.

In the process of addressing these issues we asked the question, "Can we use `green' chemistry methods to teach organic lab skills and chemical concepts using standard taper glassware on the benchtop?" Our goal was to use the latest advances in green chemistry research to develop a new lab curriculum that would decrease our reliance on fume hoods, produce a less hazardous wastestream, and make lab safer. The fact that green methods improve safety by eliminating hazards as opposed to preventing exposure through often-fallible safety controls/procedures made it ideal for implementing in a teaching lab. As we began to develop the idea, other benefits became evident:

(a) green chemical methods achieve hazard reduction at all reaction scales;

(b) incorporation of green methods is an excellent opportunity to upgrade the organic lab curriculum (that, with the exception of scale reduction, has remained virtually unchanged throughout the last century); and

(c) the curriculum provides a unique context for more detailed discussion of chemical hazards and the effects of chemicals on human health and the environment.

- Reduces laboratory waste and hazards.
- Illustrates green chemical concepts.
- Teaches modern reaction chemistry and techniques, in parallel with lecture course.
- Provides a platform for discussion of environmental issues in the classroom.
- Can be accomplished by students during a typical organic laboratory period.
- Uses inexpensive, greener solvents and reagents.
- Is adaptable to either macroscale or microscale methods.

In cases where experiments were to be modified, we adopted a set of principles to guide the process of modifying the experiments.

- Eliminate (minimize) hazardous solvents whenever possible – this includes both as reaction media and in solvent-dependent separations.
- Identify and use the most benign reagents and solvents possible – in many cases traditional reagents are too reactive, so use milder, more selective reagents.
- Develop and use efficient reaction chemistry – including catalysis, atom economy, etc.

In the process of testing the curriculum, we discovered that there are a number of unexpected benefits. One such benefit is that students are regularly required to evaluate the chemical hazards associated with a method and learn to take appropriate precautions (as opposed to taking the stance that all chemicals are hazardous and should be avoided). They learn the types of chemical hazards and how those hazards are quantified, the relevant modes of exposure or release of chemicals, and the methods to be used for safe handling of chemicals with different hazards. By addressing chemical hazards in this way, students learn to assess hazard rather than thinking that all chemicals are "toxic". This alters their (often) negative perception of chemistry.

Another significant benefit derives from students being exposed to the strategies and methods we used to make the experiments green. It would be possible to adopt a greener organic lab curriculum without involving students in the process, however, to do so is to miss a rare opportunity to change the way that students view chemistry and their perceived ability to contribute to it. For each experiment, we show the students how the new green method differs from the traditionally-used method and describe the thought process we or others used in modifying the experiment. This repeated exposure to the process of analyzing reaction conditions and finding alternative methods teaches students to:

(a) assess the conditions for hazards and likelihood of exposure/release;

(b) identify new methods that allow reduction of the hazards;

(c) explore how alternative methods can affect the course and rate of reactions; and

(d) assess the broader impact of the overall process on safety and the environment.

This aspect of the curriculum is especially empowering to students. They realize that armed with this knowledge they can help disseminate better and more sustainable chemical practices throughout academic and industrial settings acting as ambassadors of green chemistry.

3

Supercritical Carbon Dioxide

Supercritical carbon dioxide refers to carbon dioxide that is in a fluid state while also being at or above both its critical temperature and pressure, yielding rather uncommon properties. Carbon dioxide usually behaves as a gas in air at STP or as a solid called dry ice when frozen. If the temperature and pressure are both increased from STP to be at or above the critical point for carbon dioxide, it can adopt properties midway between a gas and a liquid. More specifically, it behaves as a supercritical fluid above its critical temperature (31.1°C) and critical pressure (72.9 atm), expanding to fill its container like a gas but with a density like that of a liquid. Supercritical CO_2 is becoming an important commercial and industrial solvent due to its role in chemical extraction in addition to its low toxicity and environmental impact. The relatively low temperature of the process and the stability of CO_2 also allows most compounds to be extracted with little damage or denaturing.

Uses

Carbon dioxide is gaining popularity amongst coffee manufacturers looking to move away from some of the classic decaffeinating solvents of the past, many of which led to public outcry because of real or perceived dangers related to their use in food preparation. Supercritical CO_2 is forced through the green coffee beans and then they are sprayed with water at high pressure to remove the caffeine. The caffeine can then be isolated for resale (e.g. to the pharmaceutical industry or to beverage manufacturers) by passing the water through activated charcoal filters or by distillation, crystallization or reverse osmosis.

Supramics, environmentally beneficial, low-cost substitutes for rigid thermoplastic and fired ceramic, are made using supercritical carbon dioxide as a chemical reagent. The supercritical carbon dioxide in these processes is reacted with the alkaline components of fully hardened hydraulic cement or gypsum plaster to form various carbonates. The sole by-product is ultra-pure water. Because supramics consume and sequester carbon as stable compounds in useful products, they may serve to reduce carbon that would otherwise be released into the environment.

There is considerable work being done to develop an enhanced version of a gas-turbine power production cycle to operate at temperatures near 550°C. This is a significant usage, which could have large implications for bulk thermal and nuclear generation of electricity, because the supercritical properties of carbon dioxide at above 500°C and 20 MPa enable very high thermal efficiencies, approaching 45 percent. This could increase the electrical power produced per unit of fuel required by 40 percent or more. Given the huge volume of extremely polluting fuels used in producing electricity, the potential environmental impact of such an efficient cycle could be very large.

Processes which use supercritical carbon dioxide to produce micro and nano scale particles, often for pharmaceutical uses, are currently being developed. The gas antisolvent process, rapid expansion of supercritical solutions, and supercritical antisolvent precipitation (as well as several related methods) have been shown to process a variety of substances into particles.

Supercritical carbon dioxide is also used in the foaming of polymers. Many corporations utilize supercritical carbon dioxide to saturate the polymer with solvent (carbon dioxide). Upon depressurization and heating the carbon dioxide rapidly expands, causing voids within the polymer matrix, i.e. creating a foam. Research is also ongoing at many universities in the production of microcellular foams using supercritical carbon dioxide.

Supercritical carbon dioxide is beginning to be used to enhance oil recovery in mature oil fields. At the same time, there is the possibility of using the various "clean coal" technologies which are emerging to combine such enhanced recovery methods with carbon sequestation efforts. Using gasifiers instead of conventional furnaces, coal and water is reduced to hydrogen gas, carbon dioxide, and ash. This hydrogen gas can be used to produce electrical power in combined-cycle gas turbines, while the CO_2 is captured, compressed to the supercritical state, and injected into geological storage, possibly into existing oil fields to improve yields. The unique properties of supercritical CO_2 ensure that it will remain out of the atmosphere.

Supercritical carbon dioxide is also an important emerging natural refrigerant, being used in new, low carbon solutions for domestic heat pumps These systems are undergoing continuous development with supercritical carbon dioxide heat pumps already being successfully marketed in Asia. The EcoCute systems from Japan, developed by consortium of companies including Mitsubishi, develop high temperature domestic water with small inputs of electric power by moving heat into the system from their surroundings. Their success makes a future use in other world regions possible.

In laboratories, supercritical carbon dioxide is used as an extraction solvent e.g. in determination of Total Recoverable Hydrocarbons from soils, sediments, fly-ash, and other media, and determination of PAHs in soil and solid wastes. Supercritical fluid extraction has also been used in determination of hydrocarbon components in water.

Environmental Impact

Supercritical carbon dioxide is seen as a promising green solvent because it is non-toxic, and a byproduct of other industrial processes. Furthermore, separation of the reaction components from the starting material is much simpler than with traditional organic solvents.

The Cool War

The Cool War refers to the debate about the next-generation refrigerant in car air conditioning worldwide, with an ongoing dispute between the Alliance for CO_2 Solutions supporting the uptake of sustainable CO_2 Technology in passenger cars, and chemical giants developing new chemical blends.

The Alliance and its supporters – scientists, NGOs and business leaders – urge the car industry to replace high global warming chemical substances with the natural refrigerant carbon dioxide (CO_2, R744/R-744) in car cooling and heating. This, they argue, would lead to 10% less car emissions, and knock out 1% of total greenhouse gas emissions worldwide. If CO_2 Technology is applied in other sectors, such as commercial and industrial refrigeration, heat pumps for water heating etc., it may even save up to 3% of the world's greenhouse gases.

Opponents of the Alliance claim that CO_2 Technology is not cost-efficient and safe, hence seeking to postpone the global industry decision to be taken this summer to develop new chemical blends instead.

Background

The Cool War has emanated from the decision of the European Union to phase out the current high global warming refrigerant HFC-134a in car air conditioning from January 2011 onwards To comply with the legislation, carmakers have to decide today on a new refrigerant, as they typically need 3-4 years to develop and introduce a new car platform including the air conditioning system. (It would take much less time to design merely a new air conditioning system to install in a new version of an existing model.) The current total value of the car air conditioning market is estimated to be $14.5 billion in 2007.

Arguments for CO_2

The Alliance for CO_2 Solutions and its supporters agree that the refrigerant CO_2 is:

- *More Environmental Friendly:* with the lowest Global warming potential (GWP) of all currently used and proposed refrigerants. CO_2 does not deplete the ozone layer. Since the carbon dioxide used in car air conditioning is a recycled industrial waste product it becomes environmentally neutral. Overall, using a CO_2-based air conditioning system will reduce total car emissions by 10%, thereby sparing the planet 1% of total greenhouse gases.
- *More Technically Ready:* CO_2 models have been developed and tested in all climates, being now ready for mass production. They are faster to heat and cool a car, and show a superior performance in over 90% of all driving conditions.
- *More Cost-efficient:* As a refrigerant itself, CO_2 is cheap and worldwide available. The servicing of CO_2 systems

will be less costly and less complicated than that for present systems. For the consumer the total cost of ownership is lowest with CO_2 as it will significantly cut fuel consumption by the air conditioning device. Carmakers have to make an initial investment estimated at 20 per unit, with no additional costs once CO_2 Technology enters into mass production.

- *Use as Heat Pumps:* At least one CO_2 system under development can act as a heat pump, supplying cabin heat and windshield defrosting even before the engine has warmed up.

  Natural Refrigerants such as the greenpeace-developed 'Greenfreeze', based on purified butane/propane mixtures, are entirely 'natural', and due to increased efficiency over refrigerants such as · R134a, allow the use of very small amounts of refrigerant to be used.

- *Reduced Environmental Damage:* Use of pure hydrocarbon refrigerants, which are 'backward compatible' with even early Freon (R-12) car air conditioning systems, would allow these systems to be easily converted (without modification), increasing their efficiency, and preventing further release of harmful R-134a and R-12 to the atmosphere.

Arguments Against non-CO_2 Refrigerants

Butane and propane are very flammable petroleum products; they are used as fuels for gas barbecue grills, disposable lighters, etc. (Propane has even been used as a fuel for alternative energy vehicles, often built by converting stock gasoline-engine vehicles. Like gasoline, to which it chemically is closely related, propane has a tendency to explode if ignited in an enclosed container, and an internal combustion engine designed to run on gasoline can be converted to run on propane without major modification.

The use of highly flammable hydrocarbon gases such as butane and propane as automotive refrigerants raises serious safety concerns. The EPA, in evaluating motor vehicle air conditioning substitutes for CFC-12 (Freon, or R-12) under its SNAP program, has classified as "Unacceptable Substitutes" other "Flammable blend of hydrocarbons" by reason of "insufficient data to demonstrate safety." The EPA defines "Unacceptable" in this context as "illegal for use as a CFC-12 substitute in motor vehicle air conditioners". All of the refrigerants which EPA approved for motor vehicle use in place of CFC-12 (as of September 28, 2006) contain no more than 4% total flammable hydrocarbons (butane, isobutane, and/or isopentane). Therefore, it appears unlikely, for safety reasons, that EPA will approve 'Greenfreeze' or similar hydrocarbon-based refrigerants for automotive use.

Supercritical Fluid

A supercritical fluid is any substance at a temperature and pressure above its critical point. It can diffuse through solids like a gas, and dissolve materials like a liquid. Additionally, close to the critical point, small changes in pressure or temperature result in large changes in density, allowing many properties to be "tuned". Supercritical fluids are suitable as a substitute for organic solvents in a range of industrial and laboratory processes. Carbon dioxide and water are the most commonly used supercritical fluids, being used for decaffeination and power generation respectively.

There is no surface tension in a supercritical fluid, as there is no liquid/gas phase boundary. By changing the pressure and temperature of the fluid, the properties can be "tuned" to be more liquid or more gas like. One of the most important properties is the solubility of material in the fluid. Solubility in a supercritical fluid tends to increase with density of the fluid (at constant temperature). Since density increases with pressure, then solubility also tends to increase with pressure. The relationship with temperature is a little more complicated. At constant density, solubility will increase with temperature. However, close to the critical point, the density can drop sharply with a slight increase in temperature. Therefore, close to the critical temperature, solubility often drops with increasing temperature, then rises again.

All supercritical fluids are completely miscible with each other so for a mixture a single phase can be guaranteed if the critical point of the mixture is exceeded. The critical point of a binary mixture can be estimated as the arithmetic mean of the critical temperatures and pressures of the two components,

$T_{c(mix)}$ = (mole fraction *A*) x *TcA* + (mole fraction *B*) x T_cB

For greater accuracy, the critical point can be calculated using equations of state, such as the Peng Robinson, or group contribution methods. Other properties, such as density, can also be calculated using equations of state.

Phase Diagram

Figs. 3.1 and 3.2 show projections of a phase diagram. In the pressure-temperature phase diagram the boiling separates the gas and liquid region and ends in the critical point, where the liquid and gas phases disappear to become a single supercritical phase. This can be observed in the density-pressure phase diagram for carbon dioxide, as shown in Fig. 3.2. At well below the critical temperature, e.g. 280K, as the pressure increases, the gas compresses and eventually (at just over 40 bar) condenses into a much denser liquid, resulting in the discontinuity in the line (vertical dotted line). The system consists of 2 phases in equilibrium, a dense liquid and a low density gas. As the critical temperature is approached (300K), the density of the gas at equilibrium becomes denser, and that of the liquid lower. At the critical point, (304.1 K and 7.38 MPa (73.8 bar)). there is no difference in density, and the 2 phases become one fluid phase. Thus, above the critical temperature a gas cannot be liquefied by pressure. At slightly above the critical temperature (310K), in the vicinity of the critical pressure, the line is almost vertical. A small increase in pressure causes a large increase in the density of the supercritical phase. Many other physical properties also show large gradients with pressure near the critical point, e.g. viscosity, the relative permittivity and the solvent strength, which are all closely related to the density. At higher temperatures, the fluid starts to behave like a gas, as can be seen in Fig. 3.2. For carbon dioxide at 400 K, the density increases almost linearly with pressure.

Many pressurised gases are actually supercritical fluids. For example, nitrogen has a critical point of 126.2K (- 147 °C) and 3.4 MPa (34 bar). Therefore, nitrogen in a gas cylinder above this pressure (or compressed air) is actually a supercritical fluid. These are more often known as permanent gases. At room temperature, they are well above their critical temperature, and therefore behave as a gas, similar to CO_2 at 400K above. However, they cannot be liquified by pressure unless cooled below their critical temperature.

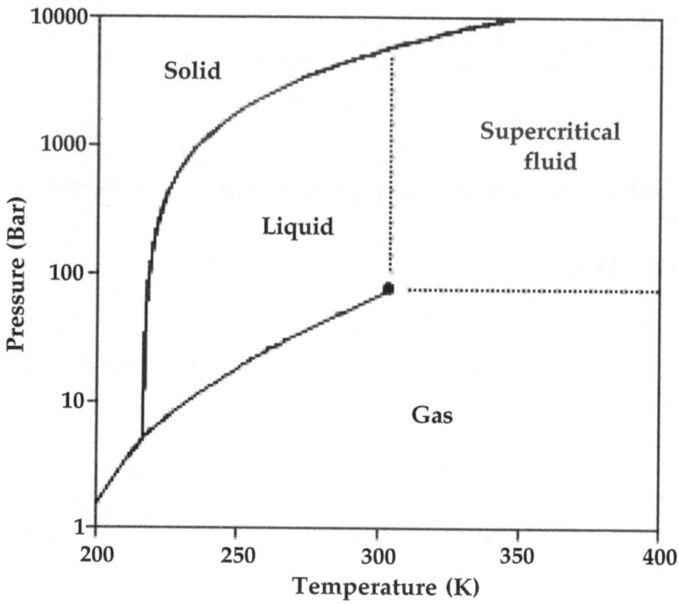

Fig. 3.1

Carbon dioxide pressure-temperature phase diagram

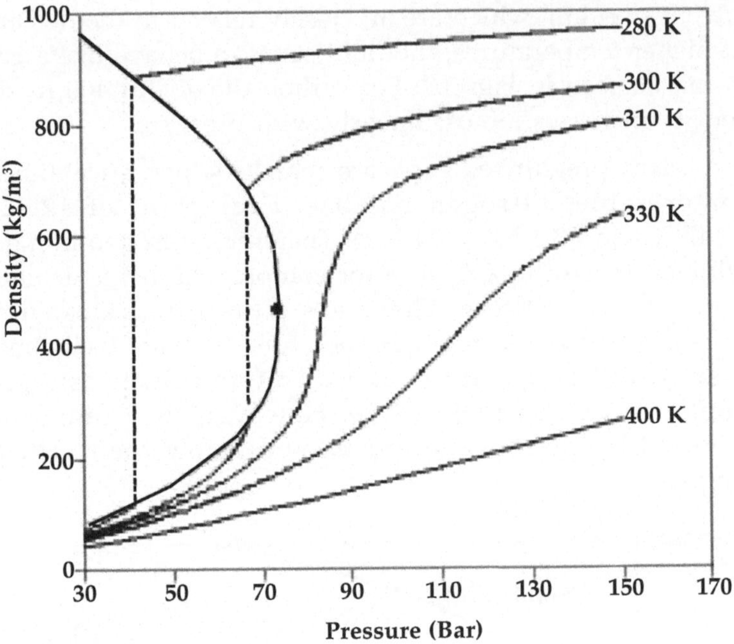

Fig. 3.2

Carbon dioxide density-pressure phase diagram

Natural Occurrence

Submarine Volcanoes

Submarine volcanoes are common features on the ocean floor. Some are active and, in shallow water, disclose their presence by blasting steam and rocky debris high above the surface of the sea. Many others lie at such great depths that the tremendous pressure from the weight of the water above them prevents the explosive release of steam and gases. This causes the water to be heated to over 375° C, turning the water in the hottest parts of the vents into a supercritical fluid since the pressure at this depth of over 3km is over 300 atmospheres, well above the 218 atmospheres required.

Venus

The atmosphere of Venus is 96.5% carbon dioxide and 3.5% nitrogen. The surface pressure is 9.3 MPa (93 bar) and the surface temperature is 735 K, above the critical points of both major constituents and making the surface atmosphere a supercritical fluid.

Applications

Supercritical Fluid Extraction

The advantages of supercritical fluid extraction (compared with liquid extraction) are that it is relatively rapid because of the low viscosities and high diffusivities associated with supercritical fluids. The extraction can be selective to some extent by controlling the density of the medium and the extracted material is easily recovered by simply depressurising, allowing the supercritical fluid to return to gas phase and evaporate leaving no or little solvent residues. Carbon dioxide is the most common supercritical solvent. It is used on a large scale for the decaffeination of green coffee beans, the extraction of hops for beer production, and the production of essential oils and pharmaceutical products from plants. Few laboratory test methods include the use of supercritical fluid extraction as an extraction method instead of using traditional solvents.

Dry Cleaning

Uses supercritical carbon dioxide instead of PERC (perchloroethylene) or other undesirable solvents for dry cleaning. Detergents which are soluble in carbon dioxide improve the solvating power of the solvent

Supercritical Fluid Chromatography

Supercritical fluid chromatography (SFC) can be used on an analytical scale, where it combines many of the advantages of HPLC and GC. It can be used with non-volatile and thermally labile analytes (unlike GC) and can be used with the universal flame ionization detector (unlike HPLC), as well as producing narrower peaks due to rapid diffusion. In practice, the advantages offered by SFC have not been sufficient to displace the widely used HPLC and GC, except in a few cases such as chiral separations and analysis of high molecular weight hydrocarbons. For manufacturing, efficient preparative simulated moving bed units are available. The purity of the final products is very high, but the cost makes it suitable only for very high value materials such as pharmaceuticals.

Chemical Reactions

Changing the conditions of the reaction solvent can allow separation of phases for product removal, or single phase for reaction. Rapid diffusion accelerates diffusion controlled reactions. Temperature and pressure can tune the reaction down preferred pathways, e.g. to improve yield of a particular chiral isomer There are also significant environmental benefits over conventional organic solvents.

Impregnation and Dyeing

Impregnation is essentially the converse of extraction. A substance is dissolved in the supercritical fluid, the solution flowed past a solid substrate, and is deposited on or dissolves in the substrate. Dyeing, which is readily carried out on polymer fibres such as polyester using disperse (non-ionic) dyes, is a special case of this. Carbon dioxide also dissolves in many polymers, considerably swelling and plasticising them and further accelerating the diffusion process.

Nano and Micro Particle Formation

The formation of small particles of a substance with a narrow size distribution is an important process in the pharmaceutical and other industries. Supercritical fluids provide a number of ways of achieving this by rapidly exceeding the saturation point of a solute by dilution, depressurization or a combination of these. These processes occur faster in supercritical fluids than in liquids, promoting nucleation or spinodal decomposition over crystal growth and yielding very small and regularly sized particles.

Supercritical drying is a method of removing solvent without surface tension effects. As a liquid dries, the surface tension_drags on small structures within a solid, causing distortion and shrinkage. Under supercritical conditions there is no surface tension, and the supercritical fluid can be removed without distortion. Supercritical drying is used for manufacture of aerogels and drying of delicate materials such as archeological samples and biological samples for electron microscopy.

Supercritical Water Oxidation

Supercritical water oxidation uses supercritical water to oxidise hazardous waste, eliminating production of toxic combustion products which burning can produce.

Supercritical Water Power Generation

The efficiency of a heat engine is ultimately dependent on the temperature difference between heat source and sink (carnot cycle). To improve efficiency of power stations the operating temperature must be raised. Using water as the coolant, this takes it into supercritical conditions. Efficiencies can be raised from about 39% for sub critical operation to about 45% using current technology. Supercritical water reactors (SCWRs) are promising advanced nuclear systems that offer similar thermal efficiency gains. Carbon dioxide can also be used in supercritical cycle nuclear plants, with similar efficiency gains.

Biodiesel Production

Conversion of vegetable oil to biodiesel is via a transesterification reaction, where the triglyceride is converted to the methyl ester plus glycerol. This is usually done using methanol and caustic or acid catalysts, but can be achieved using supercritical methanol without a catalyst.

Supercritical methanol method for biodiesel production was first studied by Saka and his coworkers. This has the advantage of allowing a greater range and water content of feedstocks (particularly used cooking oil), the product does not need to be washed to remove catalyst, and is easier to design as a continuous process.

Supramics

Supramics, environmentally beneficial, low-cost substitutes for rigid thermoplastic and fired ceramic, are made using supercritical carbon dioxide as a chemical reagent. The supercritical carbon dioxide in these processes is reacted with the alkaline components of fully hardened hydraulic cement or gypsum plaster to form various carbonates. The sole by-product is ultra-pure water. Because supramics consume and sequester carbon as stable compounds in useful products, they may serve to reduce carbon that would otherwise be released into the environment.

Carbon Capture and Storage and Enhanced Oil Recovery

Supercritical carbon dioxide is used to enhance oil recovery in mature oil fields. At the same time, there is the possibility of using "clean coal technology" to combine enhanced recovery methods with carbon sequestration. The CO_2 is separated from other flue gases either pre or post combustion, compressed to the supercritical state, and injected into geological storage, possibly into existing oil fields to improve yields.

Currently, only schemes isolating fossil CO_2 from natural gas actually use carbon storage, (e.g. Sleipner gas field) but there are many plans for future CCS schemes involving pre- or post- combustion CO_2. There is also the possibility to reduce the amount of CO_2 in the atmosphere by using biomass to generate power and sequestering the CO_2 produced.

Refrigeration

Supercritical carbon dioxide is also an important emerging natural refrigerant, being used in new, low carbon solutions for domestic heat pumps. These systems are undergoing continuous development with supercritical carbon dioxide heat pumps already being successfully marketed in Asia. The EcoCute systems from Japan, developed by consortium of companies including Mitsubishi, develop high temperature domestic water with small inputs of electric power by moving heat into the system from their surroundings. Their success makes a future use in other world regions possible

Supercritical Fluid Deposition

Suprercritical fluids can be used to deposit functional nanostructured films and nanometer-sized particles of metals onto surfaces. The gas-like surface tension, diffusivities, and viscosities allows access to nano pores much smaller than can be accessed by liquids, and the liquid-like solubilities allow much higher precursor concentrations than are typical in chemical vapour deposition This is crucial in developing more powerful electronic components, and metal particles deposited in this way are also powerful catalysts for chemical synthesis and electrochemical reactions.

History

In 1822, Baron Charles Cagniard de la Tour discovered the critical point of a substance in his famous cannon barrel experiments. Listening to discontinuities in the sound of a rolling flint ball in a sealed cannon filled with fluids at various temperatures, he observed the critical temperature. Above this temperature, the densities of the liquid and gas phases become equal and the distinction between them disappears, resulting in a single supercritical fluid phase.

4

Caffeine

Caffeine is a bitter, white crystalline xanthine alkaloid that acts as a psychoactive stimulant drug and a mild diuretic. Caffeine was discovered by a German chemist, Friedrich Ferdinand Runge, in 1819. He coined the term "kaffein", a chemical compound in coffee, which in English became caffeine. Caffeine is also part of the chemical mixtures and insoluble complexes guaranine found in guarana, mateine found in mate, and theine found in tea; all of which contain additional alkaloids such as the cardiac stimulants theophylline and theobromine, and often other chemicals such as polyphenols which can form insoluble complexes with caffeine.

Caffeine is found in varying quantities in the beans, leaves, and fruit of some plants, where it acts as a natural pesticide that paralyzes and kills certain insects feeding on the plants. It is most commonly consumed by humans in infusions extracted from the cherries of the coffee plant and the leaves of the tea bush, as well as from various foods and drinks containing products derived from the kola nut. Other sources include yerba mate, guarana berries, and the Yaupon Holly.

In humans, caffeine is a central nervous system (CNS) stimulant, having the effect of temporarily warding off drowsiness and restoring alertness. Beverages containing caffeine, such as coffee, tea, soft drinks and energy drinks enjoy great popularity. Caffeine is the world's most widely consumed psychoactive substance, but unlike many other psychoactive substances it is legal and unregulated in nearly all jurisdictions. In North America, 90% of adults consume caffeine daily The U.S. Food and Drug Administration lists caffeine as a "Multiple Purpose Generally Recognized as Safe Food Substance".

Caffeine is found in many plant species, where it acts as a natural pesticide, with high caffeine levels being reported in seedlings that are still developing foliages, but are lacking mechanical protection caffeine paralyzes and kills certain insects feeding upon the plant. High caffeine levels have also been found in the surrounding soil of coffee bean seedlings. It is therefore understood that caffeine has a natural function as both a natural pesticide and as an inhibitor of seed germination of other nearby coffee seedlings thus giving it a better chance of survival.

The most commonly used caffeine-containing plants are coffee, tea, and to a lesser extent cocoa Less commonly used sources of caffeine include the yerba maté and guarana plants, which are sometimes used in the preparation of teas and energy drinks. Two of caffeine's alternative names, *mateine* and *guaranine,* are derived from the names of these plants. Some yerba mate enthusiasts assert that mateine is a stereoisomer of caffeine, which would make it a different substance altogether. This is not true because caffeine is an achiral molecule, and therefore, has no enantiomers; nor does it have other stereoisomers. The disparity in experience and effects between the various natural caffeine sources could be due to the fact that plant sources of caffeine also contain widely varying mixtures of other xanthine alkaloids, including the cardiac stimulants theophylline and theobromine and other substances such as polyphenols which can form insoluble complexes with caffeine.

The world's primary source of caffeine is the coffee "bean" (which is actually the seed of the coffee plant), from which coffee is brewed. Caffeine content in coffee varies widely depending on the type of coffee bean and the method of preparation used; even beans within a given bush can show variations in concentration. In general, one serving of coffee ranges from 40 milligrams, for a single shot (30 millilitres) of *arabica*-variety espresso, to about 100 milligrams for a cup (120 millilitres) of drip coffee. Generally, dark-roast coffee has less caffeine than lighter roasts because the roasting process reduces the bean's caffeine content *Arabica* coffee normally contains less caffeine than the *robusta* variety Coffee also contains trace amounts of theophylline, but no theobromine.

Tea is another common source of caffeine. Although tea contains more caffeine than coffee, a typical serving contains much less, as tea is normally brewed much weaker. Besides strength of the brew, growing conditions, processing techniques and other variables also affect caffeine content. Certain types of tea may contain somewhat more caffeine than other teas. Tea contains small amounts of theobromine and slightly higher levels of theophylline than coffee. Preparation and many other factors have a significant impact on tea, and colour is a very poor indicator of caffeine content. Tea like the pale Japanese green tea gyokuro, for example, contain far more caffeine than much darker teas like lapsang souchong, which has very little.

Caffeine is also a common ingredient of soft drinks such as cola, originally prepared from kola nuts. Soft drinks typically contain about 10 to 50 milligrams of caffeine per serving. By contrast, energy drinks such as Red Bull can start at 80 milligrams of caffeine per serving. The caffeine in these drinks either originates from the ingredients used or is an additive derived from the product of decaffeination or from chemical synthesis. Guarana, a prime ingredient of energy drinks, contains large amounts of caffeine with small amounts of theobromine and theophylline in a naturally occurring slow-release excipient.

Chocolate derived from cocoa contains a small amount of caffeine. The weak stimulant effect of chocolate may be due to a combination of theobromine and theophylline as well as caffeine. Chocolate contains too little of these compounds for a reasonable serving to create effects in humans that are on par with coffee. A typical 28-gram serving of a milk chocolate bar has about as much caffeine as a cup of *decaffeinated* coffee.

In recent years various manufacturers have begun putting caffeine into shower products such as shampoo and soap, claiming that caffeine can be absorbed through the skin However, the effectiveness of such products has not been proven, and they are likely to have little stimulatory effect on the central nervous system because caffeine is not readily absorbed through the skin. nufacturers market caffeine tablets, claiming that using caffeine of pharmaceutical quality improves mental alertness. These effects have been borne out by research that shows that caffeine use (whether in tablet form or not) results in decreased fatigue and increased attentiveness These tablets are used by students studying for their exams and by people who work or drive for long hours.

History

Humans have consumed caffeine since the Stone Age.[26] Early peoples found that chewing the seeds, bark, or leaves of certain plants had the effects of easing fatigue, stimulating awareness, and elevating one's mood. Only much later was it found that the effect of caffeine was increased by steeping such plants in hot water. Many cultures have legends that attribute the discovery of such plants to people living many thousands of years ago.

According to one popular Chinese legend, the Emperor of China Shennong, reputed to have reigned in about 3000 BC, accidentally discovered that when some leaves fell into boiling water, a fragrant and restorative drink resulted Shennong is also mentioned in Lu Yu's *Cha Jing*, a famous early work on the subject of tea The history of coffee has been recorded as far back as the ninth century. During that time, coffee beans were available only in their native habitat, Ethiopia. A popular legend traces its discovery to a goatherder named Kaldi, who apparently observed goats that became elated and sleepless at night after browsing on coffee shrubs and, upon trying the berries that the goats had been eating, experienced the same vitality. The earliest literary mention of coffee may be a reference to Bunchum in the works of the 9th century Persian physician al-Razi. In 1587, Malaye Jaziri compiled a work tracing the history and legal controversies of coffee, entitled "Undat al safwa fi hill al-qahwa". In this work, Jaziri recorded that one Sheikh, Jamal-al-Din al-Dhabhani, mufti of Aden, was the first to adopt the use of coffee in 1454, and that in the 15th century the Sufis of Yemen routinely used coffee to stay awake during prayers.

Towards the close of the 16th century, the use of coffee was recorded by a European resident in Egypt, and about this time it came into general use in the Near East. The appreciation of coffee as a beverage in Europe, where it was first known as "Arabian piss," dates from the 17th century. During this time "coffee houses" were established, the first being opened in Constantinople and Venice. In Britain, the first coffee houses were opened in London in 1652, at St Michael's Alley, Cornhill. They soon became popular throughout Western Europe, and played a significant role in social relations in the 17th and 18th centuries.

The kola nut, like the coffee berry and tea leaf, appears to have ancient origins. It is chewed in many West African cultures, individually or in a social setting, to restore vitality and ease hunger pangs. In 1911, kola became the focus of one of the earliest documented health scares when the US government seized 40 barrels and 20 kegs of Coca-Cola syrup in Chattanooga, Tennessee, alleging that the caffeine in its drink was "injurious to health" On March 13, 1911, the government initiated *The United States v. Forty Barrels and Twenty Kegs of Coca-Cola,* hoping to force Coca-Cola to remove caffeine from its formula by making claims, such as that the excessive use of Coca-Cola at one girls' school led to "wild nocturnal freaks, violations of college rules and female proprieties, and even immoralities." Although the judge ruled in favor of Coca-Cola, two bills were introduced to the U.S. House of Representatives in 1912 to amend the Pure Food and Drug Act, adding caffeine to the list of "habit-forming" and "deleterious" substances which must be listed on a product's label.

The earliest evidence of cocoa use comes from residue found in an ancient Mayan pot dated to 600 BC. In the New World, chocolate was consumed in a bitter and spicy drink called *xocoatl,* often seasoned with vanilla, chile pepper, and achiote. Xocoatl was believed to fight fatigue, a belief that is probably attributable to the theobromine and caffeine content. Chocolate was an important luxury good throughout pre-Columbian Mesoamerica, and cocoa beans were often used as currency.

Chocolate was introduced to Europe by the Spaniards and became a popular beverage by 1700. They also introduced the cacao tree into the West Indies and the Philippines. It was used in alchemical processes, where it was known as Black Bean.

The leaves and stems of the Yaupon Holly (Ilex vomitoria) were used by Native Americans to brew a tea called Asi or the Black Drink the use of which among Native American groups archaeologists have demonstrated to stretch back far into antiquity, possibly dating to Late Archaic times.

Synthesis and Properties

In 1819, the German chemist Friedrich Ferdinand Runge isolated relatively pure caffeine for the first time. According to Runge, he did this at the behest of Johann Wolfgang von Goethe In 1827, Oudry isolated "theine" from tea, but it was later proved by Mulder and Jobat that theine was the same as caffeine The structure of caffeine was elucidated near the end of the 19th century by Hermann Emil Fischer, who was also the first to achieve its total synthesis This was part of the work for which Fischer was awarded the Nobel Prize in 1902. The nitrogen atoms are all essentially planar (in sp^2 orbital hybridisation), resulting in the caffeine molecule having aromatic character. Being readily available as a byproduct of decaffeination, caffeine is not usually synthesized. If desired, it may be synthesized from dimethylurea and malonic acid.

Pharmacology

Global consumption of caffeine has been estimated at 120,000 tonnes per annum making it one of the world's most popular psychoactive substances. This number equates to one serving of a caffeine beverage for every person, per day. Caffeine is a central nervous system and metabolic stimulant and is used both recreationally and medically to reduce physical fatigue and restore mental alertness when unusual weakness or drowsiness occurs. Caffeine stimulates the central nervous system first at the higher levels, resulting in increased alertness and wakefulness, faster and clearer flow of thought, increased focus, and better general body coordination, and later at the spinal cord level at higher doses. Once inside the body, it has a complex chemistry, and acts through several mechanisms as described below.

Metabolism and half-life

Caffeine from coffee or other beverages is absorbed by the stomach and small intestine within 45 minutes of ingestion and then distributed throughout all tissues of the body It is eliminated by first-order kinetics Caffeine can also be ingested rectally, evidenced by the formulation of suppositories of ergotamine tartrate and caffeine (for the relief of migraine) and chlorobutanol and caffeine (for the treatment of hyperemesis).

The half-life of caffeine—the time required for the body to eliminate one-half of the total amount of caffeine—varies widely among individuals according to such factors as age, liver function, pregnancy, some concurrent medications, and the level of enzymes in the liver needed for caffeine metabolism. In healthy adults, caffeine's half-life is approximately 3-4 hours. In women taking oral contraceptives this is increased to 5-10 hours and in pregnant women the half-life is roughly 9-11 hours Caffeine can accumulate in individuals with severe liver disease, increasing its half-life up to 96 hours In infants and young children, the half-life may be longer than in adults; half-life in a newborn baby may be as long as 30 hours. Other factors such as smoking can shorten caffeine's half-life.

Caffeine is metabolized in the liver by the cytochrome P450 oxidase enzyme system (specifically, the 1A2 isozyme) into three metabolic dimethylxanthines, which each have their own effects on the body (Fig. 4.1):

- *Paraxanthine (80%):* Has the effect of increasing lipolysis, leading to elevated glycerol and free fatty acid levels in the blood plasma.
- *Theobromine (10%):* Dilates blood vessels and increases urine volume. Theobromine is also the principal alkaloid in cocoa, and therefore chocolate.
- *Theophylline (4%):* Relaxes smooth muscles of the bronchi, and is used to treat asthma. The therapeutic dose of theophylline, however, is many times greater than the levels attained from caffeine metabolism.

Each of these metabolites is further metabolized and then excreted in the urine.

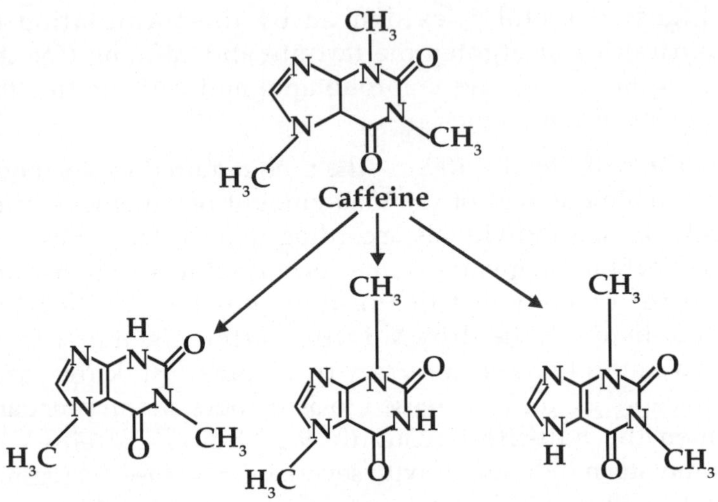

Paraxanthine (84%) Theobromine (12%) Theophylline (4%)

Fig. 4.1

Caffeine is metabolized the liver into three primary metabolites: paraxanthine (84%), theobromine (12%) and theophylline (4%)

Mechanism of Action

Like alcohol and nicotine, caffeine readily crosses the blood–brain barrier that separates the bloodstream from the interior of the brain. Once in the brain, the principal mode of action is as an antagonist of adenosine receptors. The caffeine molecule is structurally similar to adenosine, and binds to adenosine receptors on the surface of cells without activating them (an "antagonist" mechanism of action). Therefore, caffeine acts as a competitive inhibitor.

Adenosine is found in every part of the body, because it plays a role in the fundamental ATP-related energy metabolism, but it has special functions in the brain. There is a great deal of evidence that concentrations of brain adenosine are increased by various types of metabolic stress including anoxia and ischemia. The evidence also indicates that brain adenosine acts to protect the brain by suppressing neural activity and by increasing blood flow Thus, caffeine, by counteracting adenosine, has a generally disinhibitory effect on brain activity. It has not been shown, however, how these effects cause increases in arousal and alertness.

Adenosine is released in the brain through a complex mechanism. There is evidence that adenosine functions as a synaptically released neurotransmitter in some cases, but stress-related adenosine increases appear to be produced mainly by extracellular metabolism of ATP. It is not likely that adenosine is the primary neurotransmitter for any group of neurons, but rather that it is released together with other transmitters by a number of neuron types. Unlike most neurotransmitters, adenosine does not seem to be packaged into vesicles that are released in a voltage-controlled manner, but the possibility of such a mechanism has not been completely ruled out.

Several classes of adenosine receptors have been described, with different anatomical distributions. A1 receptors are widely distributed, and act to inhibit calcium uptake. A_{2A} receptors are heavily concentrated in the basal ganglia, an area that plays a critical role in behavior control, but can be found in other parts of the brain as well, in lower densities. There is evidence that A_{2A} receptors interact with the dopamine system, which is involved in reward and arousal. (A_{2A} receptors can also be found on arterial walls and blood cell membranes.)

Beyond its general neuroprotective effects, there are reasons to believe that adenosine may be more specifically involved in control of the sleep-wake cycle. McCarley and his colleagues have argued that accumulation of adenosine may be a primary cause of the sensation of sleepiness that follows prolonged mental activity, and that the effects may be mediated both by inhibition of wake-promoting neurons via A1 receptors, and activation of sleep-promoting neurons via indirect effects on A_{2A} receptors. More recent studies have provided additional evidence for the importance of A_{2A}, but not A_1, receptors.

Some of the secondary effects of caffeine are probably caused by actions unrelated to adenosine. Caffeine is known to be a competitive inhibitor of the enzyme cAMP-phosphodiesterase (cAMP-PDE), which converts cyclic AMP (cAMP) in cells to its noncyclic form, thus allowing cAMP to build up in cells. Cyclic AMP participates in activation of protein kinase A (PKA) to begin the phosphorylation of specific enzymes used in glucose synthesis. By blocking its removal caffeine intensifies and prolongs the effects of epinephrine and epine-phrine-like drugs such as amphetamine, methamphetamine, or methylphenidate. Increased concentrations of cAMP in parietal cells causes an increased activation of protein kinase A (PKA) which in turn increases activation of H^+/K^+ ATPase, resulting finally in increased gastric acid secretion by the cell. Cyclic AMP also increases the activity of the funny current, which directly increases heart rate.

Metabolites of caffeine also contribute to caffeine's effects. Paraxanthine is responsible for an increase in the lipolysis process, which releases glycerol and fatty acids into the blood to be used as a source of fuel by the muscles. Theobromine is a vasodilator that increases the amount of oxygen and nutrient flow to the brain and muscles. Theophylline acts as a smooth muscle relaxant that chiefly affects bronchioles and acts as a chronotrope and inotrope that increases heart rate and efficiency.

Effects when taken in Moderation

The precise amount of caffeine necessary to produce effects varies from person to person depending on body size and degree of tolerance to caffeine. It takes less than an hour for caffeine to begin affecting the body and a mild dose wears off in three to four hours. Consumption of caffeine does not eliminate the need for sleep: it only temporarily reduces the sensation of being tired.

With these effects, caffeine is an ergogenic: increasing the capacity for mental or physical labor. A study conducted in 1979 showed a 7% increase in distance cycled over a period of two hours in subjects who consumed caffeine compared to control subjects Other studies attained much more dramatic results; one particular study of trained runners showed a 44% increase in "race-pace" endurance, as well as a 51% increase in cycling endurance, after a dosage of 9 milligrams of caffeine per kilogram of body weight. Additional studies have reported similar effects. Another study found 5.5 milligrams of caffeine per kilogram of body mass resulted in subjects cycling 29% longer during high intensity circuits.

Caffeine citrate has proven to be of short and long term benefit in treating the breathing disorders of apnea of prematurity and bronchopulmonary dysplasia in premature infants. The only short term risk associated with caffeine citrate treatment is a temporary reduction in weight gain during the therapy and longer term studies (18 to 21 months) have shown lasting benefits of treatment of premature infants with caffeine

While relatively safe for humans, caffeine is considerably more toxic to some other animals such as dogs, horses, and parrots due to a much poorer ability to metabolize this compound. Caffeine has a much more significant effect on spiders, for example, than most other drugs do.

Caffeine relaxes the internal anal sphincter muscles and thus should be avoided by those with fecal incontinence.

Tolerance and Withdrawal

Because caffeine is primarily an antagonist of the central nervous system's receptors for the neurotransmitter adenosine, the bodies of individuals who regularly consume caffeine adapt to the continual presence of the drug by substantially increasing the number of adenosine receptors in the central nervous system. This increase in the number of the adenosine receptors makes the body much more sensitive to adenosine, with two primary consequences. First, the stimulatory effects of caffeine are substantially reduced, a phenomenon known as a tolerance adaptation. Second, because these adaptive responses to caffeine make individuals much more sensitive to adenosine, a reduction in caffeine intake will effectively increase the normal physiological effects of adenosine, resulting in unwelcome withdrawal symptoms in tolerant users. arch questions the idea that up-regulation of adenosine receptors is responsible for tolerance to the locomotor stimulant effects of caffeine, noting, among other things, that this tolerance is insurmountable by higher doses of caffeine (it should be surmountable if tolerance was due to an increase in receptors), and that the increase in adenosine receptor number is modest and does not explain the large tolerance which develops to caffeine.

Caffeine tolerance develops very quickly, especially among heavy coffee and energy drink consumers. Complete tolerance to sleep disruption effects of caffeine develops after consuming 400 mg of caffeine 3 times a day for 7 days. Complete tolerance to subjective effects of caffeine was observed to develop after consuming 300 mg 3 times per day for 18 days, and possibly even earlier. In another experiment, complete tolerance of caffeine was observed when the subject consumed 750-1200 mg per day while incomplete tolerance to caffeine has been observed in those that consume more average doses of caffeine.

Because adenosine, in part, serves to regulate blood pressure by causing vasodilation, the increased effects of adenosine due to caffeine withdrawal cause the blood vessels of the head to dilate, leading to an excess of blood in the head and causing a headache and nausea. Reduced catecholamine activity may cause feelings of fatigue and drowsiness. A reduction in serotonin levels when caffeine use is stopped can cause anxiety, irritability, inability to concentrate and diminished motivation to initiate or to complete daily tasks; in extreme cases it may cause mild depression. Together, these effects have come to be known as a "crash".

Withdrawal symptoms—possibly including headache, irritability, an inability to concentrate, drowsiness, insomnia and pain in the stomach, upper body, and joints —may appear within 12 to 24 hours after discontinuation of caffeine intake, peak at roughly 48 hours, and usually last from one to five days, representing the time required for the number of adenosine receptors in the brain to revert to "normal" levels, uninfluenced by caffeine consumption. Analgesics, such as aspirin, can relieve the pain symptoms, as can a small dose of caffeine. Most effective is a combination of both an analgesic and a small amount of caffeine.

This is not the only case where caffeine increases the effectiveness of a drug. Caffeine makes pain relievers 40% more effective in relieving headaches and helps the body absorb headache medications more quickly, bringing faster relief For this reason, many over-the-counter headache drugs include caffeine in their formula. It is also used with ergotamine in the treatment of migraine and cluster headaches as well as to overcome the drowsiness caused by antihistamines.

Overuse

In large amounts, and especially over extended periods of time, caffeine can lead to a condition known as *caffeinism.*

Caffeinism usually combines caffeine dependency with a wide range of unpleasant physical and mental conditions including nervousness, irritability, anxiety, tremulousness, muscle twitching (hyperreflexia), insomnia, headaches, respiratory alkalosis, and heart palpitations. Furthermore, because caffeine increases the production of stomach acid, high usage over time can lead to peptic ulcers, erosive esophagitis, and gastroesophageal reflux disease

There are four caffeine-induced psychiatric disorders recognized by the *Diagnostic and Statistical Manual of Mental Disorders, Fourth Edition*: caffeine intoxication, caffeine-induced anxiety disorder, caffeine-induced sleep disorder, and caffeine-related disorder not otherwise specified (NOS).

Caffeine Intoxication

An acute overdose of caffeine, usually in excess of about 300 milligrams, dependent on body weight and level of caffeine tolerance, can result in a state of central nervous system over-stimulation called *caffeine intoxication* colloquially "caffeine jitters". The symptoms of caffeine intoxication are not unlike overdoses of other stimulants. It may include restlessness, nervousness, excitement, insomnia, flushing of the face, increased urination, gastrointestinal disturbance, muscle twitching, a rambling flow of thought and speech, irritability, irregular or rapid heart beat, and psychomotor agitation In cases of much larger overdoses mania, depression, lapses in judgment, disorientation, disinhibition, delusions, hallucinations and psychosis may occur, and rhabdomyolysis (breakdown of skeletal muscle tissue) can be provoked.

In cases of extreme overdose, death can result. The median lethal dose (LD_{50}) given orally, is 192 milligrams per kilogram in rats. The LD_{50} of caffeine in humans is dependent on weight and individual sensitivity and estimated to be about 150 to 200 milligrams per kilogram of body mass, roughly 80 to 100 cups of coffee for an average adult taken within a limited time frame that is dependent on half-life.

Though achieving lethal dose with caffeine would be exceptionally difficult with regular coffee, there have been reported deaths from overdosing on caffeine pills, with serious symptoms of overdose requiring hospitalization occurring from as little as two grams of caffeine. An exception to this would be taking a drug such as Luvox which blocks the liver enzyme responsible for the metabolism of caffeine, thus increasing the central effects and blood concentrations of caffeine dramatically at five-fold. It is not contraindicated, but highly advisable to minimize the intake of caffeinated beverages, as drinking one cup of coffee will have the same effect as drinking five Death typically occurs due to ventricular fibrillation brought about by effects of caffeine on the cardiovascular system.

Treatment of severe caffeine intoxication is generally supportive, providing treatment of the immediate symptoms, but if the patient has very high serum levels of caffeine then peritoneal dialysis, hemodialysis, or hemofiltration may be required.

Anxiety and Sleep Disorders

Two infrequently diagnosed caffeine-induced disorders that are recognized by the American Psychiatric Association (APA) are *caffeine-induced sleep disorder* and *caffeine-induced anxiety disorder*, which can result from long-term excessive caffeine intake.

In the case of caffeine-induced sleep disorder, an individual regularly ingests high doses of caffeine sufficient to induce a significant disturbance in his or her sleep, sufficiently severe to warrant clinical attention.

In some individuals, the large amounts of caffeine can induce anxiety severe enough to necessitate clinical attention. This caffeine-induced anxiety disorder can take many forms, from generalized anxiety to panic attacks, obsessive-compulsive symptoms, or even phobic symptoms Because this condition can mimic organic mental disorders, such as panic disorder, generalized anxiety disorder, bipolar disorder, or even schizophrenia, a number of medical professionals believe caffeine-intoxicated people are routinely misdiagnosed and unnecessarily medicated when the treatment for caffeine-induced psychosis would simply be to stop further caffeine intake A study in the *British Journal of Addiction* concluded that caffeinism, although infrequently diagnosed, may afflict as many as one person in ten of the population.

Parkinson's Disease

Several large studies have shown that caffeine intake is associated with a reduced risk of developing Parkinson's disease (PD) in men, but studies in women have been inconclusive The mechanism by which caffeine affects PD remains a mystery. In animal models, researchers have shown that caffeine can prevent the loss of dopamine-producing nerve cells seen in Parkinson's Disease, but researchers still do not know how this occurs.

Effects on Memory and Learning

An array of studies found that caffeine could have nootropic effects, inducing certain changes in memory and learning. However, the tests performed contradict one another and the results have proven inconsistent and inconclusive.

Researchers have found that long-term consumption of low dose caffeine slowed hippocampus-dependent learning and impaired long-term memory in mice. Caffeine consumption for 4 weeks also significantly reduced hippocampal neurogenesis compared to controls during the experiment. The conclusion was that long-term consumption of caffeine could inhibit hippocampus-dependent learning and memory partially through inhibition of hippocampal neurogenesis.

In another study, caffeine was added to rat neurons *in vitro*. The dendritic spines (a part of the brain cell used in forming connections between neurons) taken from the hippocampus (a part of the brain associated with memory) grew by 33% and new spines formed. After an hour or two, however, these cells returned to their original shape.

Another study showed that subjects—after receiving 100 milligrams of caffeine—had increased activity in brain regions located in the frontal lobe, where a part of the working memory network is located, and the anterior cingulate cortex, a part of the brain that controls attention. The caffeinated subjects also performed better on the memory tasks.

However, a different study showed that caffeine could impair short term memory and increase the likelihood of the tip of the tongue phenomenon. The study allowed the researchers to suggest that caffeine could aid short-term memory when the information to be recalled is related to the current train of thought, but also to hypothesize that caffeine hinders short-term memory when the train of thought is unrelated In essence, caffeine consumption increases mental performance related to focused thought while it may decrease broad-range thinking abilities.

Effects on the Heart

Caffeine binds to receptors on the surface of heart muscle cells which leads to an increase in the level of cAMP inside the cells (by blocking the enzyme that degrades cAMP), mimicking the effects of epinephrine (which binds to receptors on the cell that activate cAMP production). cAMP acts as a "second messenger," and activates a large number of protein kinase A (PKA; cAMP-dependent protein kinase). This has the overall effect of increasing the rate of glycolysis and increases the amount of ATP available for muscle contraction and relaxation. According to one study, caffeine in the form of coffee, significantly reduces the risk of heart disease in epidemiological studies. However, the protective effect was found only in participants who were not severely hypertensive (i.e. patients that are not suffering from a very high blood pressure). Furthermore, no significant protective effect was found in participants aged less than 65 years or in cerebrovascular disease mortality for those aged equal or more than 65 years.

Effects on Children

Scientific studies contradict the common belief that caffeine consumption causes stunted growth in children. Children also can experience the same effects from caffeine as adults. Most energy drinks (containing extremely high amounts of caffeine) have been banned in many schools throughout the world.

Water Extraction

Coffee beans are soaked in water. The water, which contains not only caffeine but also many other compounds which contribute to the flavor of coffee, is then passed through activated charcoal, which removes the caffeine. The water can then be put back with the beans and evaporated dry, leaving decaffeinated coffee with a good flavor. Coffee manufacturers recover the caffeine and resell it for use in soft drinks and over-the-counter caffeine tablets.

Supercritical Carbon Dioxide Extraction

Supercritical carbon dioxide is an excellent nonpolar solvent for caffeine, and is safer than the organic solvents that are otherwise used. The extraction process is simple: CO_2 is forced through the green coffee beans at temperatures above 31.1 °C and pressures above 73 atm. Under these conditions, CO_2 is in a "supercritical" state: it has gaslike properties which allow it to penetrate deep into the beans but also liquid-like properties which dissolve 97–99% of the caffeine. The caffeine-laden CO_2 is then sprayed with high pressure water to remove the caffeine. The caffeine can then be isolated by charcoal adsorption (as above) or by distillation, recrystallization, or reverse osmosis.

Extraction by Organic Solvents

Organic solvents such as ethyl acetate present much less health and environmental hazard than previously used chlorinated and aromatic solvents. Another method is to use triglyceride oils obtained from spent coffee grounds.

Religion

Some Latter-day Saints (Mormons), Seventh-day Adventists, Church of God (Restoration) adherents, and Christian Scientists do not consume caffeine. A few followers from these religions believe that God is opposed to the use of all non-medical psychoactive substances.

The Church of Jesus Christ of Latter-day Saints has said the following with regard to caffeinated beverages: "With reference to cola drinks, the Church has never officially taken a position on this matter, but the leaders of the Church have advised, and we do now specifically advise, against the use of any drink containing harmful habit-forming drugs under circumstances that would result in acquiring the habit. Any beverage that contains ingredients harmful to the body should be avoided."

Gaudiya Vaishnava Hindus generally also abstain from caffeine, as it is alleged to cloud the mind and over-stimulate the senses. To be initiated under a guru, one must have had no caffeine (along with alcohol, nicotine and other drugs) for at least a year.

5

Hydrogen Peroxide

Introduction

Hydrogen peroxide was first isolated in 1818 by Louis Jacques Thénard by reacting barium peroxide with nitric acid An improved version of this process used hydrochloric acid, followed by sulfuric acid to precipitate the barium sulfate byproduct. Thénard's process was used from the end of the 19th century until the middle of the 20th century.

For a long time it was believed that pure hydrogen peroxide was unstable, because attempts to separate the hydrogen peroxide from the water, which is present during synthesis, failed. This was because traces of solids and heavy metal ions led to a catalytic decomposition or explosions of the hydrogen peroxide. 100% pure hydrogen peroxide was first obtained through vacuum distillation by Richard Wolffenstein in 1894 At the end of 19th century, Petre Melikishvili and his pupil L. Pizarjevski showed that of the many proposed formulas of hydrogen peroxide, the correct one was H-O-O-H.

Hydrogen peroxide (H_2O_2) is a very pale blue liquid which appears colorless in a dilute solution, slightly more viscous than water. It is a weak acid. It has strong oxidizing properties and is therefore a powerful bleaching agent that is mostly used for bleaching paper, but has also found use as a disinfectant, as an oxidizer, as an antiseptic, and in rocketry (particularly in high concentrations as high-test peroxide or HTP) as a monopropellant, and in bipropellant systems The oxidizing capacity of hydrogen peroxide is so strong that the chemical is considered a highly reactive oxygen species.

Hydrogen peroxide is naturally produced as a byproduct of oxygen metabolism, and virtually all organisms possess enzymes known as peroxidases, which harmlessly and catalytically decompose low concentrations of hydrogen peroxide to water and oxygen.

Structure and Properties

As with all molecules, the physical properties of hydrogen peroxide are the result of its molecular mass, structure and distribution of atoms within the molecule.

Molecular Structure

The preferred molecular structure of any molecule is the configuration which has the lowest internal stress. For hydrogen peroxide, there are two basic structural forms (conformers) available for the molecule. While flat shape of the *anti* conformer would minimize steric repulsions, the 90° torsion angle of the *syn* conformer would optimize mixing between the filled *p-type* orbital of the oxygen (one of the lone pairs) and the LUMO of the vicinal O-H bond.

The resulting anti-clinal "skewed" shape is a compromise between the two conformers.

Despite the fact that the O-O bond is a single bond, the molecule has a remarkably high barrier to complete rotation of 29.45 kJ/mol (compared with 12.5 kJ/mol for the rotational barrier of ethane). The increased barrier is attributed to repulsion between one lone pair and other lone pairs. The bond angles are affected by hydrogen bonding, which is relevant to the structural difference between gaseous and crystalline forms; indeed a wide range of values is seen in crystals containing molecular H_2O_2.

Comparison with Analogues

Analogues of hydrogen peroxide include the chemically identical deuterium peroxide and malodorous hydrogen disulfide. Hydrogen disulfide has a boiling point of only 70.7°C despite having a higher molecular weight, indicating that hydrogen bonding increases the boiling point of hydrogen peroxide.

Physical Properties of Hydrogen Peroxide Solutions

Aqueous hydrogen peroxide solutions have specific properties that are different from those of the pure chemical due to hydrogen bonding between water and hydrogen peroxide molecules. Specifically, hydrogen peroxide and water form a eutectic mixture, exhibiting freezing-point depression. While pure water melts and freezes at approximately 273K, and pure hydrogen peroxide just 0.4K below that, a 50% (by volume) solution melts and freezes at 221 K.

Manufacture

Hydrogen peroxide is manufactured today almost exclusively by the autoxidation of 2-ethyl-9,10-dihydroxy-anthracene ($C_{16}H_{14}O_2$) to 2-ethylanthraquinone ($C_{16}H_{12}O_2$) and hydrogen peroxide using oxygen from the air. This is known as the Riedl-Pfleiderer process.

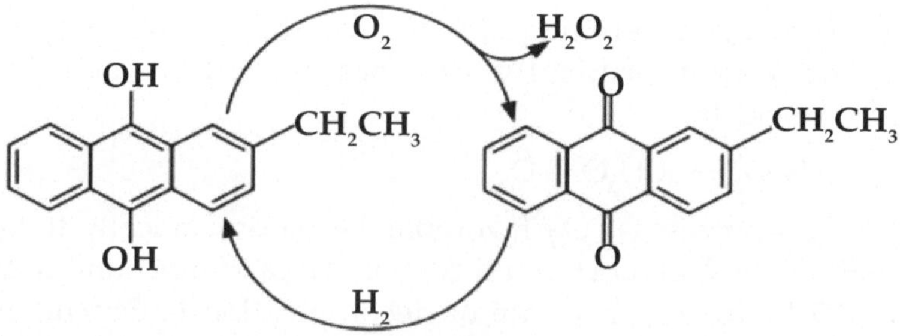

Fig. 5.1

In this reaction, the hydroxy groups on the middle ring of anthracene are deprotonated and are turned into ketones, while two double bonds are lost from the middle ring and are replaced as C=O double bonds in the ketone groups. The anthraquinone derivative is then extracted out and reduced back to the dihydroxy compound using hydrogen gas in the presence of a metal catalyst. The overall equation for the process is deceptively simple:

$$H_2 + O_2 \rightarrow H_2O_2$$

However the economics of the process depend on effective recycling of the quinone and extraction solvents, and of the hydrogenation catalyst.

Formerly inorganic processes were used, employing the electrolysis of an aqueous solution of sulfuric acid or acidic ammonium bisulfate (NH_4HSO_4), followed by hydrolysis of the peroxydisulfate $[(SO_4)_2]^{2-}$ which is formed.

In 1994, world production of H_2O_2 was around 1.9 million tonnes and grew to 2.2 million in 2006 most of which was at a concentration of 70% or less. In that year bulk 30% H_2O_2 sold for around US $0.54 per kg, equivalent to US $1.50 per kg (US $0.68 per lb) on a "100% basis".

In 2009, a new process of producing hydrogen peroxide using gold-palladium nanoparticles has been reported. This process claims to be inexpensive, efficient, and environmental friendly.

Decomposition

Hydrogen peroxide always decomposes (disproportionates) exothermically into water and oxygen gas spontaneously:

$$2\ H_2O_2 \rightarrow 2\ H_2O + O_2$$

This process is very favorable thermodynamically. It has a ΔH° of -98.2 kJ·mol^{-1} and a ΔG° of -119.2 kJ·mol^{-1} and a ΔS of 70.5 J·mol^{-1}·K^{-1}. The rate of decomposition is dependent on the temperature and concentration of the peroxide, as well as the pH and the presence of impurities and stabilizers. Hydrogen peroxide is incompatible with many substances that catalyse its decomposition, including most of the transition metals and their compounds. Common catalysts include manganese dioxide, and silver. The same reaction is catalysed by the enzyme catalase, found in the liver, whose main function in the body is the removal of toxic byproducts of metabolism and the reduction of oxidative stress. The decomposition occurs more rapidly in alkali, so acid is often added as a stabilizer.

The liberation of oxygen and energy in the decomposition has dangerous side effects. Spilling high concentrations of hydrogen peroxide on a flammable substance can cause an immediate fire, which is further fueled by the oxygen released by the decomposing hydrogen peroxide. High-strength peroxide (also called high-test peroxide, or HTP) must be stored in a suitable, vented container to prevent the buildup of oxygen gas, which would otherwise lead to the eventual rupture of the container.

In the presence of certain catalysts, such as Fe^{2+} or Ti^{3+}, the decomposition may take a different path, with free radicals such as HO· (hydroxyl) and HOO· being formed. A combination of H_2O_2 and Fe^{2+} is known as Fenton's reagent.

A common concentration for hydrogen peroxide is "20 volume", which means that when 1 volume of hydrogen peroxide is decomposed, it produces 20 volumes of oxygen. A 20 "volume" concentration of hydrogen peroxide is equivalent to 1.79 mol/dm^3 (Molar solution) or about six per cent.

Hydrogen peroxide available at drug stores is three percent solution. In such small concentrations, it is less stable, and decomposes faster. It is usually stabilized with acetanilide, a substance which has toxic side effects in significant amounts.

Redox Reactions

H_2O_2 is one of the most powerful oxidizers known — stronger than chlorine, chlorine dioxide, and potassium permanganate. Also, through catalysis, H_2O_2 can be converted into hydroxyl radicals (.OH) with reactivity second only to fluorine.

In aqueous solution, hydrogen peroxide can oxidize or reduce a variety of inorganic ions. When it acts as a reducing agent, oxygen gas is also produced. In acidic solutions Fe^{2+} is oxidized to Fe^{3+},

$$2\ Fe^{2+}(aq) + H_2O_2 + 2\ H^+(aq) \rightarrow 2\ Fe^{3+}(aq) + 2H_2O(l)$$

and sulfite (SO_3^{2-}) is oxidized to sulfate (SO_4^{2-}). However, potassium permanganate is reduced to Mn^{2+} by acidic H_2O_2. Under alkaline conditions, however, some of these reactions reverse; for example, Mn^{2+} is oxidized to Mn^{4+} (as MnO_2).

Another example of hydrogen peroxide acting as a reducing agent is the reaction with sodium hypochlorite, which is a convenient method for preparing oxygen in the laboratory.

$$NaOCl + H_2O_2 \rightarrow O_2 + NaCl + H_2O$$

Hydrogen peroxide is frequently used as an oxidizing agent in organic chemistry. One application is for the oxidation of thioethers to sulfoxides. For example, methyl phenyl sulfide was oxidised to methyl phenyl sulfoxide in

99% yield in methanol in 18 hours (or 20 minutes using a $TiCl_3$ catalyst):

$$Ph\text{-}S\text{-}CH_3 + H_2O_2 \rightarrow Ph\text{-}S(O)\text{-}CH_3 + H_2O$$

Alkaline hydrogen peroxide is used for epoxidation of electron-deficient alkenes such as acrylic acids, and also for oxidation of alkylboranes to alcohols, the second step of hydroboration-oxidation.

Formation of Peroxide Compounds

Hydrogen peroxide is a weak acid, and it can form hydroperoxide or peroxide salts or derivatives of many metals.

For example, on addition to an aqueous solution of chromic acid (CrO_3) or acidic solutions of dichromate salts, it will form an unstable blue peroxide $CrO(O_2)_2$. In aqueous solution it rapidly decomposes to form oxygen gas and chromium salts.

It can also produce peroxoanions by reaction with anions; for example, reaction with borax leads to sodium perborate, a bleach used in laundry detergents:

$$Na_2B_4O_7 + 4\ H_2O_2 + 2\ NaOH \rightarrow 2\ Na_2B_2O_4(OH)_4 + H_2O$$

H_2O_2 converts carboxylic acids (RCOOH) into peroxy acids (RCOOOH), which are themselves used as oxidizing agents. Hydrogen peroxide reacts with acetone to form acetone peroxide, and it interacts with ozone to form hydrogen trioxide, also known as trioxidane. Reaction with urea produces carbamide peroxide, used for whitening teeth. An acid-base adduct with triphenylphosphine oxide is a useful "carrier" for H_2O_2 in some reactions.

Alkalinity

Hydrogen peroxide is a much weaker base than water, but it can still form adducts with very strong acids. The superacid HF/SbF5 forms unstable compounds containing the $[H_3O_2]^+$ ion.

Uses

Industrial Applications

About 50% of the world's production of hydrogen peroxide in 1994 was used for pulp- and paper-bleaching. Other bleaching applications are becoming more important as hydrogen peroxide is seen as an environmentally benign alternative to chlorine-based bleaches. It is highly corrosive to metal.

Other major industrial applications for hydrogen peroxide include the manufacture of sodium percarbonate and sodium perborate, used as mild bleaches in laundry detergents. It is used in the production of certain organic peroxides such as dibenzoyl peroxide, used in polymerisations and other chemical processes. Hydrogen peroxide is also used in the production of epoxides such as propylene oxide. Reaction with carboxylic acids produces a corresponding peroxy acid. Peracetic acid and meta-chloroperoxybenzoic acid (commonly abbreviated mCPBA) are prepared from acetic acid and *meta*-chlorobenzoic acid, respectively. The latter is commonly reacted with alkenes to give the corresponding epoxide.

In the PCB manufacturing process, hydrogen peroxide mixed with sulfuric acid was used as the microetch chemical for copper surface roughening preparation.

A combination of a powdered precious metal-based catalyst, hydrogen peroxide, methanol and water can produce superheated steam in one to two seconds, releasing only CO_2 and high temperature steam for a variety of purposes.

There has been increased use of vaporized hydrogen peroxide in the validation and bio-decontamination of half suit and glove port isolators in pharmaceutical production.

Nuclear Pressurized Water Reactors (PWRs) use hydrogen peroxide during the plant shutdown to force the oxidation and dissolution of activated corrosion products deposited on the fuel. The corrosion products are then removed with the cleanup systems before the reactor is disassembled.

Hydrogen Peroxide is also used in the Oil and Gas exploration industry to oxidise rock matrix in preparation for micro-fossil analysis.

Domestic Uses

- Diluted H_2O_2 (between 3% and 12%) is used to bleach human hair when mixed with ammonia, hence the phrase "peroxide blonde".
- It is absorbed by skin upon contact and creates a local skin capillary embolism which appears as a temporary whitening of the skin.
- It is used to whiten bones that are to be put on display.
- The strength of a solution may be described as a percentage or volume, where 1% hydrogen peroxide releases 3.3 volumes of oxygen during decomposition. Thus, a 3% solution is equivalent to 10 volume and a 6% solution to 20 volume, etc.
- Three per cent H_2O_2 is used medically for cleaning wounds, removing dead tissue, and as an oral debriding agent. Peroxide stops slow (small vessel) wound bleeding/oozing, as well. Most over-the-counter peroxide solutions are not suitable for ingestion.
- Three per cent H_2O_2 is effective at treating fresh (red) blood-stains in clothing and on other items. It must be applied to clothing before blood stains can be accidentally "set" with heated water. Cold water and soap are then used to remove the peroxide treated blood.
- The United States Food and Drug Administration (FDA) has classified hydrogen peroxide as a Low Regulatory Priority (LRP) drug for use in controlling fungus on fish and fish eggs.

- Some horticulturalists and users of hydroponics advocate the use of weak hydrogen peroxide solution ("Spanish water") in watering solutions. Its spontaneous decomposition releases oxygen that enhances a plant's root development and helps to treat root rot (cellular root death due to lack of oxygen) and a variety of other pests. There is some peer-reviewed academic research to back up some of the claims.
- Laboratory tests conducted by fish culturists in recent years have demonstrated that common household hydrogen peroxide can be used safely to provide oxygen for small fish. Hydrogen peroxide releases oxygen by decomposition when it is exposed to catalysts such as manganese dioxide.
- Hydrogen peroxide is a strong oxidizer effective in controlling sulfide and organic related odors in wastewater collection and treatment systems. It is typically applied to a wastewater system where there is a retention time of 30 minutes to 5 hours before hydrogen sulfide is released. Hydrogen peroxide oxidizes the hydrogen sulfide and promotes bio-oxidation of organic odors. Hydrogen peroxide decomposes to oxygen and water, adding dissolved oxygen to the system thereby negating some Biochemical Oxygen Demand (BOD).
- Mixed with baking soda and a small amount of hand soap, hydrogen peroxide is effective at removing skunk odor.
- If a dog has swallowed a harmful substance (e.g. chocolate), small amounts of hydrogen peroxide can be given to induce vomiting.
- Hydrogen peroxide is used with phenyl oxalate ester and an appropriate dye in glow sticks as an oxidizing agent. It reacts with the ester to form an unstable CO_2

dimer which excites the dye to an excited state; the dye emits a photon (light) when it spontaneously relaxes back to the ground state.

- Diluted hydrogen peroxide can be used in a 50/50 mixture with white vinegar in the removal of accumulated lead and gun powder from a stainless steel fire-arm suppressor. The acetic acid is the main reagent, but the bubbling action of the hydrogen peroxide is used to agitate the mixture and accelerate the reaction. The chemical mixture should be replaced every 24 hours as it will become ineffective after that period. Responsible care of the waste mixture is required as the end result is an aqueous solution of lead acetate, which is highly toxic and must be disposed of through a Hazardous Waste facility. Alternately, the solution may be reacted with salt water or sulfuric acid to precipitate insoluble lead chloride or lead sulfate, respectively. The solids may then be removed by mechanical filtration and are safer to dispose of than aqueous lead acetate. This method should not be used on aluminum suppressors, as acetic acid will dissolve the aluminum and form aluminum acetate

Use as Propellant

H_2O_2 can be used either as a monopropellant (not mixed with fuel) or as the oxidizer component of a bipropellant rocket. Use as a monopropellant takes advantage of the decomposition of 70-98+% concentration hydrogen peroxide into steam and oxygen. The propellant is pumped into a reaction chamber where a catalyst, usually a silver or platinum screen, triggers decomposition, producing steam at over 600 °C which is expelled through a nozzle, generating thrust. H_2O_2 monopropellant produces a maximum specific impulse (I_{sp}) of 161 s (1.6 kN·s/kg), which makes it a low-performance monopropellant. Peroxide generates much less thrust than hydrazine, but is not toxic. The Bell Rocket Belt used hydrogen peroxide monopropellant.

As a bipropellant H_2O_2 is decomposed to burn a fuel as an oxidizer. Specific impulses as high as 350 s (3.5 kN·s/kg) can be achieved, depending on the fuel. Peroxide used as an oxidizer gives a somewhat lower I_{sp} than liquid oxygen, but is dense, storable, noncryogenic and can be more easily used to drive gas turbines to give high pressures using an efficient *closed cycle*. It can also be used for regenerative cooling of rocket engines. Peroxide was used very successfully as an oxidizer in World-War-II German rockets (e.g. T-Stoff for the Me-163), and for the low-cost British Black Knight and Black Arrow launchers.

In the 1940s and 1950s the Walter turbine used hydrogen peroxide for use in submarines while submerged; it was found to be too noisy and require too much maintenance compared to diesel-electric power systems. Some torpedoes used hydrogen peroxide as oxidizer or propellant, but this was dangerous and has been discontinued by most navies. Hydrogen peroxide leaks were blamed for the sinkings of HMS *Sidon* and the Russian submarine *Kursk*. It was discovered, for example, by the Japanese Navy in torpedo trials, that the concentration of H_2O_2 in right-angle bends in HTP pipework can often lead to explosions in submarines and torpedoes. SAAB Underwater Systems is manufacturing the Torpedo 2000. This torpedo, used by the Swedish navy, is powered by a piston engine propelled by HTP as an oxidizer and kerosene as a fuel in a bipropellant system.

While rarely used now as a monopropellant for large engines, small hydrogen peroxide attitude control thrusters are still in use on some satellites. They are easy to throttle, and safer to fuel and handle before launch than hydrazine thrusters. However, hydrazine is more often used in spacecraft because of its higher specific impulse and lower rate of decomposition.

Recently H_2O_2/propylene has been proposed as an approach to inexpensive Single Stage To Orbit: a fuel tank containing propylene has a bladder floating in it containing H_2O_2. This combination offers 15% superior Isp to O_2/RP4 (a kerosene used as rocket propellant), does not need turbines or cryogenic storage or hardware, and greatly reduces the cost of the booster. The potential of this and other alternative systems is discussed in some detail at Dunn Engineering.

Therapeutic Use

Hydrogen peroxide is *generally recognized as safe* (GRAS) as an antimicrobial agent, an oxidizing agent and for other purposes by the US Food and Drug Administration

Hydrogen peroxide has been used as an antiseptic and anti-bacterial agent for many years due to its oxidizing effect. While its use has decreased in recent years with the popularity of better-smelling and more readily-available over the counter products, it is still used by many hospitals, doctors and dentists in sterilizing, cleaning and treating everything from floors to root canal procedures.

- Like many oxidative antiseptics, hydrogen peroxide causes mild damage to tissue in open wounds, but it also is effective at rapidly stopping capillary bleeding (slow blood oozing from small vessels in abrasions), and is sometimes used sparingly for this purpose, as well as cleaning.
- Hydrogen peroxide can be used as a toothpaste when mixed with correct quantities of baking soda and salt.
- Hydrogen peroxide and benzoyl peroxide are sometimes used to treat acne.
- Hydrogen peroxide is used as an emetic in veterinary practice.

Alternative Uses

- The American Cancer Society states that "there is no scientific evidence that hydrogen peroxide is a safe, effective or useful cancer treatment", and advises cancer patients to "remain in the care of qualified doctors who use proven methods of treatment and approved clinical trials of promising new treatments."

- Another controversial alternative medical procedure is inhalation of hydrogen peroxide at a concentration of about 1%. Internal use of high concentration hydrogen peroxide has a history of causing fatal blood disorders, and its recent use as a therapeutic treatment has been linked to several deaths.

Safety

Regulations vary, but low concentrations, such as 3%, are widely available and legal to buy for medical use. Higher concentrations may be considered hazardous and are typically accompanied by a Material Safety Data Sheet (MSDS). In high concentrations, hydrogen peroxide is an aggressive oxidizer and will corrode many materials, including human skin. In the presence of a reducing agent, high concentrations of H_2O_2 will react violently.

Hydrogen peroxide should be stored in a cool, dry, well-ventilated area and away from any flammable or combustible substances It should be stored in a container composed of non-reactive materials such as stainless steel or glass (other materials including some plastics and aluminium alloys may also be suitable) Because it breaks down quickly when exposed to light, it should be stored in an opaque container, and pharmaceutical formulations typically come in brown bottles that filter out light.

Hydrogen peroxide, either in pure or diluted form, can pose several risks:

- *Explosive Vapours:* Above roughly 70% concentrations, hydrogen peroxide can give off vapor that can detonate above 70° C (158° F) at normal atmospheric pressure This can then cause a boiling liquid expanding vapor explosion (BLEVE) of the remaining liquid. Distillation of hydrogen peroxide at normal pressures is thus highly dangerous.

- *Hazardous Reactions:* Hydrogen peroxide vapors can form sensitive contact explosives with hydrocarbons such as greases. Hazardous reactions ranging from ignition to explosion have been reported with alcohols, ketones, carboxylic acids (particularly acetic acid), amines and phosphorus.
- *Spontaneous ignition:* Concentrated hydrogen peroxide, if spilled on clothing (or other flammable materials), will preferentially evaporate water until the concentration reaches sufficient strength, at which point the material may spontaneously ignite.
- *Corrosive:* Concentrated hydrogen peroxide (>50%) is corrosive, and even domestic-strength solutions can cause irritation to the eyes, mucous membranes and skin. Swallowing hydrogen peroxide solutions is particularly dangerous, as decomposition in the stomach releases large quantities of gas (10 times the volume of a 3% solution) leading to internal bleeding. Inhaling over 10% can cause severe pulmonary irritation.
- *Bleach agent:* Low concentrations of hydrogen peroxide, on the order of three per cent or less, will chemically bleach many types of clothing to a pinkish hue. Caution should be exercised when using common products that may contain hydrogen peroxide, such as facial cleaner or contact lens solution, which easily splatter upon other surfaces.
- *Internal Ailments:* Large oral doses of hydrogen peroxide at a 3% concentration may cause "irritation and blistering to the mouth, (which is known as Black Hairy Tongue) throat, and abdomen", as well as "abdominal pain, vomiting, and diarrhea".

- *Vapour Pressure:* Hydrogen peroxide has a significant vapor pressure (1.2 kPa at 50 °C [CRC *Handbook of Chemistry and Physics*, 76th Ed, 1995-1996]) and exposure to the vapor is potentially hazardous. Hydrogen peroxide vapor is a primary irritant, primarily affecting the eyes and respiratory system and the NIOSH Immediately dangerous to life and health limit (IDLH) is only 75 ppm. Documentation for Immediately Dangerous to Life or Health Concentrations (IDLH): Long term exposure to low ppm concentrations is also hazardous and can result in permanent lung damage and OSHA Occupational Safety and Health Administration has established a permissible exposure limit of 1.0 ppm calculated as an eight hour time weighted average (29 CFR 1910.1000, Table Z-1) and hydrogen peroxide has also been classified by the ACGIH American Conference of Industrial Hygienists (ACGIH) as a "known animal carcinogen, with unknown relevance on humans.[2008 Threshold Limit Values for Chemical Substances and Physical Agents & Biological Exposure Indices, ACGIH] In applications where high concentrations of hydrogen peroxide are used, suitable personal protective equipment should be worn and it is prudent in situations where the vapor is likely to be generated, such as hydrogen peroxide gas or vapor sterilization, to ensure that there is adequate ventilation and the vapor concentration monitored with a continuous gas monitor for hydrogen peroxide. Continuous gas monitors for hydrogen peroxide are available from several suppliers. Further information on the hazards of hydrogen peroxide is available from OSHA Occupational Safety and Health Guideline for Hydrogen Peroxide and from the ATSDR. Agency for Toxic Substances and Disease Registry.
- Vitiligo is an acquired skin disorder with the loss of native skin pigment, which affects about 0.5-1% of the world

population. Recent studies have discovered increased H_2O_2 levels in the epidermis and in blood are one of many hallmarks of this disease.

Historical Incidents

- Several people received minor injuries after a hydrogen peroxide spill on board Northwest Airlines flight 957 from Orlando to Memphis on October 28, 1998 and subsequent fire on Northwest Airlines flight 7.
- During the Second World War, Doctors in Nazi concentration camps experimented with the use of hydrogen peroxide injections in the killing of human subjects.
- Hydrogen peroxide was said to be one of the ingredients in the bombs which failed to explode in the 21 July 2005 London bombings.
- The Russian submarine K-141 Kursk sailed out to sea to perform an exercise of firing dummy torpedoes at the Pyotr Velikiy, a Kirov class battlecruiser. On August 12, 2000 at 11:28 local time (07:28 UTC), there was an explosion while preparing to fire the torpedoes. The only credible report to date is that this was due to the failure and explosion of one of the Kursk's hydrogen peroxide-fuelled torpedoes. It is believed that HTP, a form of highly concentrated hydrogen peroxide used as propellant for the torpedo, seeped through rust in the torpedo casing. A similar incident was responsible for the loss of HMS Sidon in 1955.

6

Supercritical Water Oxidation

Supercritical water oxidation or SCWO is a process that occurs in water at temperatures and pressures above a mixture's thermodynamic critical point. Under these conditions water becomes a fluid with unique properties that can be used to advantage in the destruction of hazardous wastes such as PCBs. The fluid has a density between that of water vapor and liquid at standard conditions, and exhibits high gas-like diffusion rates along with high liquid-like collision rates. In addition, solubility behavior is reversed so that chlorinated hydrocarbons become soluble in the water, allowing single-phase reaction of aqueous waste with a dissolved oxidizer. The reversed solubility also causes salts to precipitate out of solution, meaning they can be treated using conventional methods for solid-waste residuals. Efficient oxidation reactions occur at low temperature (400-650 °C) with reduced NOx production.

The SCWO can be classified as green chemistry or as a Clean Technology. The elevated pressures and temperatures required for SCWO are routinely encountered in industrial applications such as petroleum refining and chemical synthesis.

A unique addition to the world of supercritical water (SCW) oxidation is generating high-pressure flames inside the SCW medium. The pioneer works on high-pressure Supercritical Water Flames were carried out by professor EU Franck at the German University of Karlsruhe in the late 80s. The works were mainly aimed at anticipating conditions which would cause spontaneous generation of non-desirable flames in the flameless SCW oxidation process. These flames would cause instabilities to the system and its components. ETH Zurich pursued the investigation of hydrothermal flames in continuously operated reactors. The rising needs for waste treatment and destruction methods motivated a Japanese Group in the Ebara Corporation to explore SCW flames as an environmental tool.

Commercial Applications

Several companies in the United States are working to commercialize supercritical reactors to destroy hazardous wastes. Widespread commercial application of SCWO technology requires a reactor design capable of resisting fouling and corrosion under supercritical conditions.

In Japan a number of commercial SCWO applications exist, among them one unit for treatment of halogenated waste built by Organo. In Korea two commercial size units have been built by Hanwha.

In Europe, Chematur Engineering AB of Sweden commercialized the SCWO technology for treatment of spent chemical catalysts to recover the precious metal, the AquaCat process. The unit has been built for Johnson Matthey in the UK. It is the only commercial SCWO unit in Europe and with its capacity of 3 m^3/h it is the largest SCWO unit in the world. Chematur's Super Critical Fluids technology was acquired by SCFI Group (Cork, Ireland) who are actively commercializing the Aqua Critox SCWO process for treatment of sludge, e.g. deinking sludge and sewage sludge. Many long duration trials on these applications have been made and thanks to the high destruction efficiency of 99.9%+ the solid residue after the SCWO process is well suited for recycling. In the case of deinking sludge as paper filler and in the case of sewage sludge as phosphorus and coagulant. SCFI Group operate a six metric ton/day Aqua Critox demonstration plant in Cork, Ireland.

Cializing their patented transpiring wall SCWO reactor ("TWR") with a focus on renewable energy applications. The TWR has been demonstrated to provide superior resistance to fouling and corrosion. The TWR is capable of operating at reaction temperatures in excess of 800° C, at pressures above or below the critical pressure of water, and has been demonstrated at a feedrate of 0.5 metric tons/day. Turbosystems Engineering SCWO Technology.

Supercritical

Supercritical may refer to:

- *Critical mass,* the smallest amount of fissile material needed for a sustained nuclear chain reaction.
- *Critical temperature,* Tc, the temperature above which distinct liquid and gas phases do not exist.
- *Supercritical fluid,* a substance at a temperature and pressure above its thermodynamic critical point.
- *Supercritical carbon dioxide,* same as above, applied to carbon dioxide.
- *Supercritical fluid chromatography,* high performance liquid chromatography (HPLC), using supercritical carbon dioxide as the mobile phase.
- *Supercritical airfoil,* an airfoil designed to delay the onset of wave drag in the transonic speed range.

- *Supercritical drying*, the process to remove liquid in a precisely controlled way, similar to freeze drying.
- *Supercritical water reactor* (SCWR), a Generation IV nuclear reactor concept that uses supercritical water as the working fluid.
- *Supercritical water oxidation* or SCWO, a process that occurs in water at temperatures and pressures above a mixture's thermodynamic critical point.
- *Supercritical flow*, when the flow velocity is larger than the wave velocity.
- *Supercritical steam generators*, a steam generator operating above the critical point of water, hence having no water-steam separation.
- *Hopf bifurcation*, a local bifurcation ; when the first Lyapunov coefficient is negative, the bifurcation is called supercritical.
- *Pitchfork bifurcation*, similar to the case of a Hopf bifurcation.

Critical Mass

A critical mass is the smallest amount of fissile material needed for a sustained nuclear chain reaction. The critical mass of a fissionable material depends upon its nuclear properties (e.g. the nuclear fission cross-section), its density, its shape, its enrichment, its purity, its temperature, and its surroundings.

Explanation of Criticality

The term critical refers to an equilibrium fission reaction (steady-state or continuous chain reaction); this is where there is no increase or decrease in power, temperature, or neutron population.

A numerical measure of a critical mass is dependent on the neutron multiplication factor, *k*, where:

$k = f - l$

where f is the average number of neutrons released per fission event and l is the average number of neutrons lost by either leaving the system or being captured in a non-fission event. When $k = 1$ the mass is critical.

A subcritical mass is a mass of fissile material that does not have the ability to sustain a fission reaction. A population of neutrons introduced to a subcritical assembly will exponentially decrease, typically rapidly. In this case, $k < 1$. A steady rate of spontaneous fissions causes a proportional steady level of neutron activity. The constant of proportionality increases as k increases.

A *supercritical* mass is one where there is an increasing rate of fission. The material may settle into equilibrium (i.e. become critical again) at an elevated temperature/power level or destroy itself (disassembly is an equilibrium state). In the case of supercriticality, $k > 1$.

Changing the Point of Criticality

The point, and therefore the mass, where criticality occurs may be changed by modifying certain attributes, such as fuel, shape, temperature, density, and the installation of a neutron-reflective substance. These attributes have complex interactions and interdependencies, this section explains only the simplest ideal cases.

Varying the Amount of Fuel

It is possible for a fuel assembly to be critical at near zero power. If the perfect quantity of fuel were added to a slightly subcritical mass to create an "exactly critical mass", fission would be self-sustaining for one neutron generation (fuel consumption makes the assembly subcritical).

If the perfect quantity of fuel were added to a slightly subcritical mass, to create a barely supercritical mass, the temperature of the assembly would increase to an initial maximum (for example: 1 K above the ambient temperature) and then decrease back to room temperature after a period of time, because fuel consumed during fission brings the assembly back to subcriticality once again.

Changing the Shape

A mass may be exactly critical, but not a perfect homogeneous sphere. Changing the shape to be closer to a perfect sphere will make the mass supercritical. Conversely, changing the shape to be further from a sphere will decrease its reactivity, making it subcritical.

Changing the Temperature

A mass may be exactly critical at a particular temperature. Increasing the temperature of the material, (assuming an ideal material that does not expand or contract, see Varying the density of the mass) will cause the atoms to vibrate faster which causes atoms to "see" and generate neutrons with a wider spread of velocities. This is termed doppler broadening and has a significant effect on generally reducing criticality (and hence increasing the required critical mass). This is because any U238 in the mass will now see and absorb more fast neutrons, thus removing them from the chain. (This is for a simple largely unreacted enriched Uranium 235 case, in practice detailed examinations of all isotopes and lifetimes are needed to determine the actual temperature coefficient, which can also be positive.)

Varying the Density of the Mass

The higher the density, the lower the critical mass. The density of a material at a constant temperature can be changed by varying the pressure or tension or by changing crystal structure. An ideal mass will become subcritical if allowed to expand or conversely the same mass will become supercritical if compressed. Changing the temperature may also change the density, however the effect on critical mass is then complicated by the temperature effects (See Changing the temperature), and by whether the material expands or contracts with increased temperature. Assuming the material expands with temperature (enriched Uranium 235 at room temperature for example), at an exactly critical state, it will become subcritical if warmed, hence lower density, or conversely the same mass will become supercritical if cooled, hence higher density. Such a material is said to have a negative temperature coefficient of reactivity to indicate that its reactivity decreases when its temperature increases. Using such a material as fuel means fission reduces as the fuel temperature increases.

Use of a Neutron Reflector

Surrounding a spherical critical mass with a neutron reflector further reduces the mass needed for criticality. A good neutron reflector is beryllium metal. This increases the rate of neutron collisions, resulting in supercriticality.

Criticality in Nuclear Weapon Design

Until detonation is desired, a nuclear weapon must be kept sub-critical. In the case of a uranium bomb, this can be achieved by keeping the fuel in a number of separate pieces, each below the critical size either because they are too small or unfavourably shaped. To produce detonation, the uranium is brought together rapidly. In Little Boy, this was achieved by firing a piece of uranium (a 'Doughnut'), down a gun barrel onto another piece, (a 'Spike'), a design referred to as a *gun-type fission weapon*.

A theoretical 100% pure Pu-239 weapon could also be constructed as a gun-type weapon. In reality, this is impractical because even "weapons grade" Pu-239 is contaminated with a small amount of Pu-240, which has a strong propensity toward spontaneous fission. Because of this, a reasonably sized gun-type weapon would suffer nuclear reaction before the masses of plutonium would be in a position for a full-fledged explosion to occur.

Instead, the plutonium is present as a subcritical sphere (or other shape), which may or may not be hollow. Detonation is produced by exploding a shaped charge surrounding the sphere, increasing the density (and collapsing the cavity, if present) to produce a prompt critical configuration. This is known as an *implosion type weapon.*

Critical Point (Thermodynamics)

Carbon dioxide creating a fog when cooling from supercritical to critical temperature.

In physical chemistry, thermodynamics, chemistry and condensed matter physics, a critical point, also called a critical state, specifies the conditions (temperature, pressure and sometimes composition) at which a phase boundary ceases to exist. There are multiple types of critical points such as vapor-liquid critical points and liquid-liquid critical points.

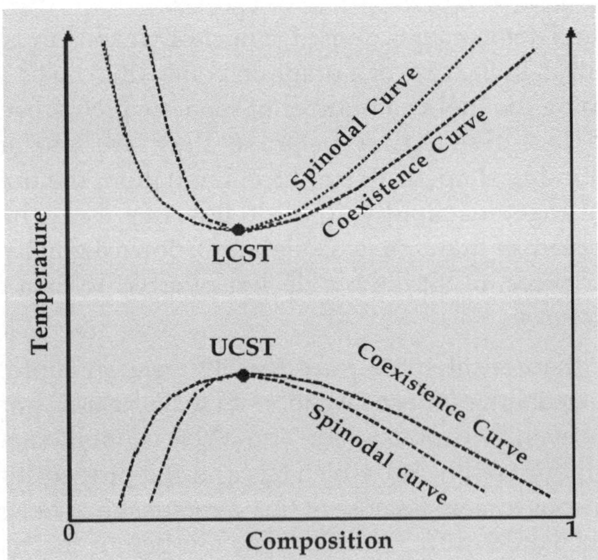

Fig 6.1

A plot of typical polymer solution phase behaviour including two critical points: an LCST and a UCST

The liquid-liquid critical point of a solution denotes the limit of the two-phase region of the phase diagram. In other words, this is the point at which an infinitesimal change in some thermodynamic variable such as temperature or pressure will lead to separation of the mixture into two distinct liquid phases, as shown in the polymer-solvent phase diagram to the right. Such points may also be denoted a *critical solution temperature.* Two types of liquid-liquid critical points are the upper critical solution temperature, or UCST, which denotes the warmest point at which cooling will induce phase separation, and the lower critical solution temperature, which denotes the coolest point at which heating will induce phase separation.

Mathematical Definition

From a theory standpoint, as seen in the figure to the right the liquid-liquid critical point represents the temperature-concentration extremum of the spinodal curve. Thus in a two-component system it must satisfy two conditions. First is the condition of the spinodal curve, which is that the second derivative of the free energy with respect to concentration must equal zero. The second condition is the extremum condition that the *third* derivative of the free energy with respect to concentration must also equal zero, or that the derivative of the spinodal temperature with respect to concentration must equal zero.

Vapour-liquid Critical Point

The term *"critical point"* is sometimes used to specifically denote the vapour-liquid critical point of a material. The vapor-liquid critical point denotes the conditions above which distinct liquid and gas phases do not exist.

As shown in the pure species phase diagram to the right, this is the point at which the phase boundary between liquid and gas terminates. In water, the critical point occurs at around 647 K (374 °C or 705 °F) and 22.064 MPa (3200 PSIA or 218atm).

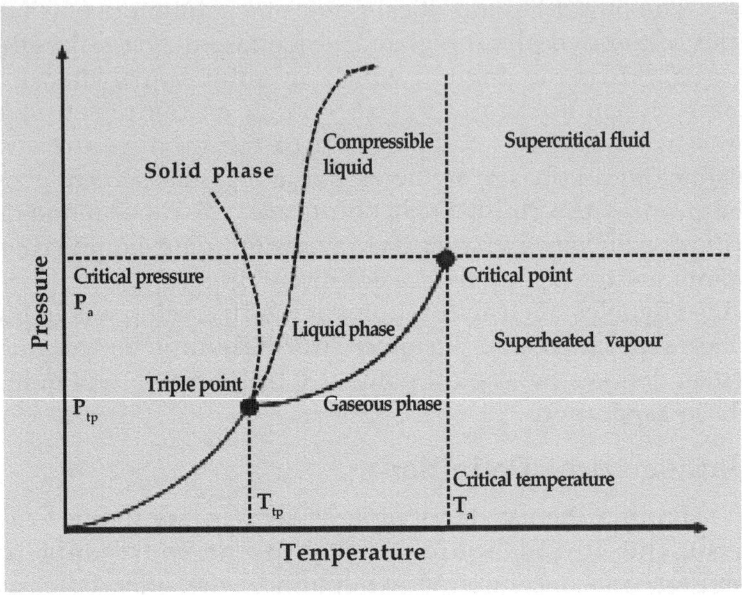

Fig. 6.2

The vapor-liquid critical point in a pressure-temperature phase diagram is at the high temperature extreme of the liquid-gas phase boundary. The dotted green line should be ignored

As the critical temperature is approached, the properties of the gas and liquid phases approach one another, at the critical point resulting in only one phase: a homogeneous supercritical fluid. The heat of vaporization is zero at and beyond this critical point, so there is no distinction between the two phases. Above the critical temperature a liquid cannot be formed by an increase in pressure, but with enough pressure a solid may be formed. The critical pressure is the vapor pressure at the critical temperature. On the diagram showing the thermodynamic properties for a given substance, the point at critical temperature and critical pressure is called the critical point of the substance. The critical molar volume is the volume of one mole of material at the critical temperature and pressure.

Critical properties vary from material to material, just as is the case for the melting point and boiling point. Critical properties for many pure substances are readily available in the literature. Obtaining critical properties for mixtures is somewhat more problematic.

Mathematical Definition

For pure substances, there is an inflection point in the critical isotherm on a pV diagram. This means that at the critical point:

$$\left(\frac{\partial p}{\partial V}\right)_T = \left(\frac{\partial^2 p}{\partial V^2}\right)_T = 0$$

This relation can be used to evaluate two parameters for an equation of state in terms of the critical properties.

Sometimes a set of reduced properties are defined in terms of the critical properties, ie.:

$$T_r = T/T_c$$

$$p_r = p/p_c$$

$$V_r = V/V_c$$

Principle of Corresponding States

Critical variables are useful for rewriting a varied equation of state into one that applies to all materials. The effect is similar to a normalizing constant. The principle of corresponding states indicates that substances at equal reduced pressures and temperatures have equal reduced volumes. This relationship is approximately true for many substances, but becomes increasingly inaccurate for large values of p_r.

In Renormalization Group Theory

According to renormalization group theory, the defining property of criticality is that the natural length scale characteristic of the structure of the physical system, the so-called correlation length?, becomes infinite. There are also lines in phase space along which this happens: these are *critical lines.*

In equilibrium systems the critical point is reached only by tuning a control parameter precisely. However, in some non-equilibrium systems the critical point is an attractor of the dynamics in a manner that is robust with respect to system parameters, a phenomenon referred to as self-organized criticality. The critical point is described by a conformal field theory.

Supercritical Airfoil

A supercritical airfoil is an airfoil designed, primarily, to delay the onset of wave drag in the transonic speed range. Supercritical airfoils are characterized by their flattened upper surface, highly cambered (curved) aft section, and greater leading edge radius as compared to traditional airfoil shapes. The supercritical airfoils were designed in the 1960s, by then NASA scientist Richard Whitcomb, and were first tested on the TF-8A Crusader. While the design was initially developed as part of the supersonic transport (SST) project at NASA, it has since been mainly applied to increase the fuel efficiency of many high subsonic aircraft. Research in 1940 by DVL's K.A. Kawalki led to subsonic profiles very similar to the supercritical profiles, which was the basis for the objection in 1984 against the US-patent specification for the supercritical airfoil. The supercritical airfoil shape is incorporated into the design of a supercritical wing.

Research aircraft of the 1950s and 60s found it difficult to break the sound barrier, or even reach Mach 0.9, with conventional airfoils. Supersonic airflow over the upper surface of the traditional airfoil induced excessive wave drag and a form of stability loss called Mach tuck. Due to the airfoil shape used, supercritical wings experience these problems less severely and at much higher speeds, thus allowing the wing to maintain high performance at speeds closer to Mach 1. Techniques learned from studies of the original supercritical airfoil sections are used to design airfoils for high-speed subsonic and transonic aircraft from the Boeing 777 to the McDonnell Douglas AV-8B Harrier II.

Supercritical airfoils have four main benefits: they have a higher drag divergence Mach number they develop shock waves farther aft than traditional airfoils, they greatly reduce shock-induced boundary layer separation, and their geometry allows for more efficient wing design (e.g., a thicker wing and/or reduced wing sweep, each of which may allow for a lighter wing). At a particular speed for a given airfoil section, the critical Mach number, flow over the upper surface of an airfoil can become locally supersonic, but slow down to match the pressure at the trailing edge of the lower surface without a shock. However, at a certain higher speed, the drag divergence Mach number, a shock is required to recover enough pressure to match the pressures at the trailing edge. This shock causes transonic wave drag, and induces flow separation behind it; both have negative effects on the airfoil's performance.

At a certain point along the airfoil, a shock is generated, which increases the pressure coefficient to the critical value Cp-crit, where the local flow velocity will be Mach 1. The position of this shockwave is determined by the geometry of the airfoil; a supercritical foil is more efficient because the shockwave is minimized and is created as far aft as possible thus reducing drag. Compared to a typical airfoil section, the supercritical airfoil creates more of its lift at the aft end, due to its more even pressure distribution over the upper surface.

In addition to improved transonic performance, a supercritical wing's enlarged leading edge gives it excellent high-lift characteristics. As a result, aircraft utilizing a supercritical wing have superior takeoff and landing performance. This makes the supercritical wing a favorite for designers of cargo transport aircraft. A notable example of one such heavy-lift aircraft that uses a supercritical wing is the C-17 Globemaster III.

Supercritical Water Reactor

The Supercritical water reactor (SCWR) is a Generation IV reactor concept that uses supercritical water as the working fluid. SCWRs are basically LWRs operating at higher pressure and temperatures with a direct, once-through cycle. As most commonly envisioned, it would operate on a direct cycle, much like a BWR, but since it uses supercritical water (not to be confused with critical mass) as the working fluid, would have only one phase present, like the PWR. It could operate at much higher temperatures and pressure than both current PWRs and BWRs.

Supercritical water-cooled reactors (SCWRs) are promising advanced nuclear systems because of their high thermal efficiency (i.e., about 45% vs. about 33% efficiency for current light water reactors (LWR) and considerable plant simplification.

A key issue in natural circulation is constituted by the stability of the flow mainly when two phase conditions are concerned and when the feedback with neutron kinetics is possible.

The main mission of the SCWR is generation of low-cost electricity. It is built upon two proven technologies, LWRs, which are the most commonly deployed power generating reactors in the world, and supercritical fossil fuel fired boilers, a large number of which are also in use around the world. The SCWR concept is being investigated by 32 organizations in 13 countries.

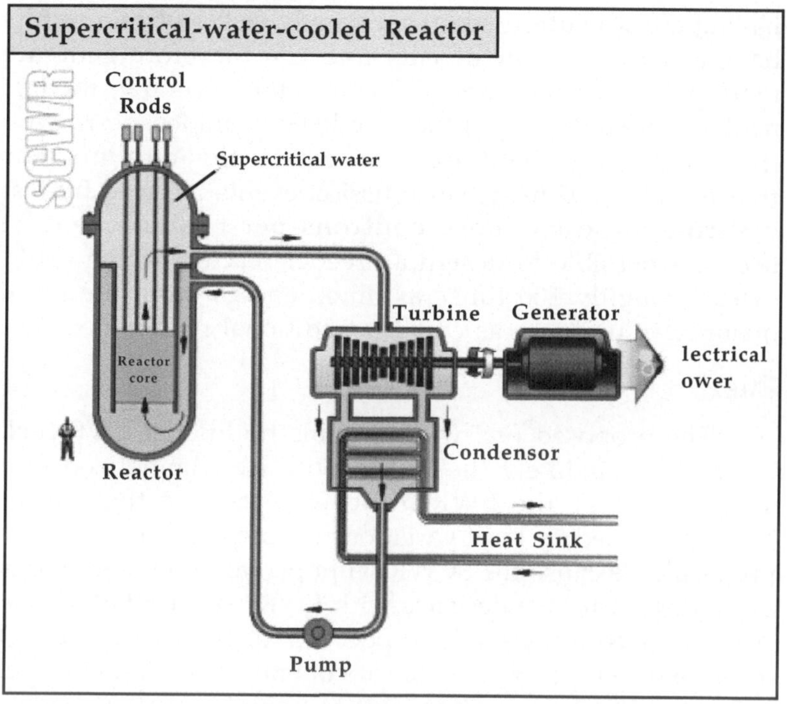

Fig. 6.3

Supercritical water reactor scheme

The reactors are intended for use in nuclear power plants to produce nuclear power from nuclear fuel.

Design

Moderator

The SCWR uses water as a neutron moderator. Moderation comes primarily from the high density sub-critical water. This high-density water is either introduced from cooling tubes inserted into the core or as a reflector or moderated-part of the core.

The neutron spectrum will be only partly moderated, perhaps to the point that the SCWR will technically become a fast neutron reactor. There are three main advantages for having a fast neutron spectrum. Firstly fast neutron reactors have a higher power density and can therefore generate more power for the same size of reactor. Secondly, the fast neutrons are able to split the long lived actinides, destroying the most long lived nuclear waste through nuclear transmutation. Thirdly, since fission events induced by fast neutrons produce more neutrons per fission event, it becomes possible to design a breeder reactor, which could extract roughly 100 times as much energy from the same quantity of uranium as could a traditional reactor design.

Fuel

The proposed fuel will resemble traditional LWR fuel. However, it is likely the SCWR will use channelized fuel assemblies like the BWR in order to reduce the risk of hotspots caused by local variations in core thermal hydraulic properties. Because the SCWR will operate under conditions exceeding current experience with LWRs and LMFBRs, new criteria for core materials (especially for the fuel cladding) must be developed to ensure safe operation to maintain fuel rod integrity during abnormal transients, normal power operation as well as concerns for release of the fission products caused by oxidation corrosion of the cladding. There are four failure modes considered for the fuel rod's integrity during abnormal transient conditions: mechanical failure, buckling collapse, over pressure damage and creep failure. It is expected that hydrogen will need to be injected into the coolant to reduce oxidation corrosion of the sheath.

Coolant

The coolant will be supercritical water. Operating above the critical pressure ensures the coolant remains single-phase in the core. At a lower pressure it would boil, producing chaotic voids (bubbles) with less density and therefore less moderating effect, making the reactor power output hard to predict and control. At extreme pressure, above the critical point, steam and liquid can be considered to be the same density, and indistinguishable. The hope is that more of the heat produced from fission can be converted into electricity in reactors cooled and/or moderated with supercritical water. Additionally, the elements handling water's phase change from liquid to gas in conventional light water reactors are not needed. This simplification should reduce construction costs and improve reliability and safety. Current LWRs need recirculation and jet pumps, pressurizers, steam generators, and steam separators and dryers, all or most of which would not be required.

Control

SCWRs would likely have control rods inserted through the top, as is done in PWRs.

Material

This section describes the R & D needs for SCWR material. The actual R&D needed to select and/or develop materials that be compatible and resist under condition above condition current LWR. the R & D proposed for SCWR designs focus on following key areas:

- Oxidation, corrosion, and stress corrosion cracking
- Radiolysis and water chemistry
- Strength, embrittlement, and creep resistance
- Dimensional and microstructural stability.

In addition to those performance factors, the cost of the material and its effect on fuel utilization must also be considered to meet the economic and sustainability requirements of Generation IV designs.

Advantages and Disadvantages

Advantages

- Supercritical water-cooled reactors promise to have thermal efficiencies of approximately 45% versus the current 33% of light water reactors.

- The supercritical coolant has a high specific enthalpy.
- The SCWR design is far simpler than current designs, eliminating circulation pumps, pressurizers, steam generators, steam separators and dryers.

Disadvantages

- Many of the materials needed for the SCWR are either expensive, rare, or do not exist.
- Specific start-up procedures required to avoid instability.
- Unknown chemistry.

7

Waste Minimization

Waste minimization is a waste management approach that focuses on reducing the amount and toxicity of hazardous waste that is generated. In addition to hazardous wasted regulated under RCRA, the EPA encourages waste minimization techniques that focus on preventing waste from ever being created, (source reduction) and recycling. There are three general methods of waste minimization: source reduction, recycling, and treatment.

Source Reduction

Changing practices and processes to reduce or eliminate the generation of hazardous wastes and materials is referred to as source reduction. Some source reduction methods include chemical substitution, process modification, and improved operating procedures. Some examples are:

1. Not mixing hazardous and non-hazardous waste streams, this helps to reduce the amount of hazardous waste to be disposed of.

2. Good housekeeping practices:
   (a) Seal and contain processes to prevent the escape of fumes or leaks to the environment.
   (b) Taking care when weighing and transferring chemicals to minimize spills.
3. Consider the use of micro scale laboratory experiments; this uses smaller amounts of chemicals.
4. Minimize your inventory and buy chemicals in quantities that will be used in the near future. A lot of hazardous chemical waste is generated when too much of a chemical is bought and having it go bad before it is used.
5. Date all chemical containers when received. Use a first-in, first-out inventory system; this will help to prevent chemicals expiring before they are used.
6. Maintain labels on all chemicals and if chemicals are transferred to secondary containers, the new container should be labeled with the chemical name; this helps to prevent the generation of unknown wastes. Unknown chemical wastes are harder and much more expensive to dispose of.
7. Chemical Substitution is when a non-hazardous chemical or a chemical of a lesser toxicity can be used in place of another. Evaluate procedures to see if less hazardous chemicals can be substituted for ones that are used. A less toxic chemical may reduce cost of disposal as well as a reduced exposure to workers.

Recycling

Another method of waste minimization is recycling. Recycling is when a waste material is used for another purpose, treated and reused in the same process, or reclaimed for another process. Some examples include:

1. Purchasing gas cylinders, including lecture bottles from manufacturers who will accept the return of the partially used or empty cylinders.

2. Photographic wastes can be treated with silver recovery units to recover the silver.
3. Collecting ballasts, equipment, and lead acid, mercury, lithium, and nickel-cadmium batteries for commercial recycling.
4. Reclaiming metallic mercury if the mercury is not mixed with any other waste streams.
5. Redistilling used solvents.
6. Recirculate unused or surplus chemicals within your department.

Treatment

The last technique for waste minimization is treatment of waste; some wastes can be treated to render them non-hazardous or less hazardous. Some examples of treatment:

1. Neutralize acids and bases.
2. Oxidize cyanide salts and ethidium bromide with bleach solutions.
3. Polymerize acrylamide solutions.
4. Inject gels directly with ethidium bromide to eliminate large volumes of liquid waste.

If treatment is not a part of the end step of an experiment and is done separately from the experiment, it is considered hazardous waste treatment, which cannot be done without a treatment permit from the Louisiana Department of Environmental Quality. At LSU Health Science Center all treatment is done by an outside licensed hazardous waste contractor.

Some treatment of waste can be done on site like neutralizing or detoxifying a chemical as the final step of an experiment. These steps could decrease or eliminate toxicity or help to reduce the volume of waste. Please contact the Environmental Health and Safety Office before initializing any treatment procedures of hazardous waste.

Why is Waste Minimization Important?

Waste minimization is important because it helps to protect the environment and it makes good business sense. Waste minimization saves money through avoided disposal costs, creates safer working conditions for employees, and protects human health and the environment.

Conclusion

All departments that generate waste at LSU Health Sciences Center should make waste minimization an active and on going component of their operations. This means that individual laboratories need to take responsibility for their by products of operations and the waste that is generated. Because the actual generators are most familiar with their work and the materials they use, they are the best source for new ideas on how to minimize waste. For waste minimization to be effective, all generators of hazardous waste need to have an active role. Waste minimization is a feasible and obtainable goal if carefully planned and implemented.

Types of Chemical Reactions

Doing chemical reactions is the main focus of chemistry because that's where the action is. That's when something gets made, changed, or destroyed. So a lot of emphasis is placed on chemical reactions.

Chemical reactions can be understood by these three areas of focus. A reaction is basically shuffling around the building blocks of chemistry (atoms/ions/molecules/electrons). Force and energy is what decides if the reaction occurs and how fast. Mathematics helps you keep an inventory of all the starting and ending materials.

As mentioned, chemical reactions mostly focuses on the building blocks. More specifically, how they are put together, how they dissassemble, and ways to rearrange them.

Chemical reactions are often classified into these five categories. Let's go through them one by one.

Synthesis is the making of some compound such as a drug, a flavoring, a plastic, a jewel.

If atoms were big enough to pick up, we could easily make all kinds of interesting materials. Of course, they are extremely small and it's not that easy, but the concept is just as easy.

The synthesis reaction is normally represented by the letters "A" and "B" forming a compound represented by "AB". "A" and "B" could represent just two different elements, or they could be compounds (2 or more elements). (roll cursor over image for animation)

Flavorings (Example of Synthesis)

Many artificial flavorings are made from combining an alcohol with an organic acid. An alcohol has an "OH" group. An organic acid has a carboxylic acid group (COOH). The questions marks means there is more attached to these groups. Depending what is attached, a different flavor is synthesized.

When the alcohol and organic acid combine, water is released. Below you will see some of the reactions that make the flavors you see here.

Decomposition Reactions

We are interested in decomposition reactions for two different reasons. Sometimes we want to speed up decomposition and other times we want to slow down decomposition.

Here's an example of decomposition of turkey waste. We don't eat turkey feathers, heads, feets, and guts, so there's a need to get rid of this waste. Some companies have figured out ways to break down all of this waste into an oil. Remember most tissue (muscles, protein, etc.) in animals are made with very long chains of carbon. The process is to break down these long chains to about 10 to 20 carbon atoms long. That makes an oil similar to diesel or biodiesel.

The general formula for decomposition is shown here. This could mean two atoms that are bonded fly apart, or it can mean a larger molecule with multiple atoms coming apart. (roll cursor over image to see animation).

Potassium perchlorate is shown. When it decomposes it releases oxygen. That oxygen is used to speed up the combustion of material it's in contact with. For example, potassium perchlorate is used in fireworks. Other material such as paper or sawdust mixed with it will explode because of this source of oxygen.

Here's a fun decomposition reaction. The starting compound is ammonium dichromate. When heated, it begins to decompose into nitrogen gas, water vapor and powdered chromium (III) oxide. It looks like a volcano with ash being spread all over the place.

This looks like the compound is burning, but looking at the chemical equation we see no oxygen is being consumed. So the flame it produces does not need oxygen. So you can't put out this kind of "fire" by smothering it like most fires.

Here is ammonium dichromate again but this time showing a molecule of ammonium dichromate decomposing. (roll cursor over image to see animation).

Whenever you see gases in the products realize that the reaction can be dangerous because it can build up pressure and cause an explosion.

Single Replacement

Another type of reaction is called single replacement (or single displacement). One element that is not bonded to any other element (A) comes in to take the place of another element that is bonded to something (B). Element (B) is now freed.

8

Safer Chemicals

EU chemicals policy must ensure a high level of protection of human health and the environment as enshrined in the Treaty both for the present generation and future generations while also ensuring the ing of the internal market and the competitiveness of the chemical industry. Fundamental to achieving these objectives is the Precautionary Principle. Whenever reliable scientific evidence is available that a substance may have an adverse impact on human health and the environment but there is still scientific uncertainty about the precise nature or the magnitude of the potential damage, decision-making must be based on precaution in order to prevent damage to human health and the environment.

Another important objective is to encourage the substitution of dangerous by less dangerous substances where suitable alternatives are available. It is also essential to ensure the efficient functioning of the internal market and the competitiveness of the chemical industry.

Many synthetic chemicals are extremely useful and bring great benefits to our lives and our health. But many are also extremely dangerous. discriminately exposed to chemical pollutants on an ongoing basis, but we are. Research into levels of industrial chemicals in the human body shows that we are continuously exposed to a large number of chemical pollutants The fact that we are all continuously exposed to many different chemicals is because the law allows this to happen. European law is currently based on an assumption that there are 'acceptable' levels of exposure, even to the most hazardous chemicals. Regulators determine acceptable levels of risk from these exposures.

It is further assumed that the quantity of substances that we are exposed to can be controlled through dilution and dispersion of chemicals throughout the environment. However, this assumption falls apart for chemicals which do not degrade, or degrade only slowly in the environment and which can bioaccumulate. Additionally, a surprising number of hazardous chemicals are used in consumer products and there is another, implicit, assumption that exposure to hazardous chemicals from consumer products and other dispersed sources, is negligible. Recent research suggests it is not. As a result of these assumptions we are all continuously, and quite legally, exposed to multiple and ongoing small doses of many different substances.

As long as chemical regulation is based on this risk-based philosophy, human and environmental exposure to dangerous chemicals – 'chemicals of very high concern' – will continue. The 'disperse and dilute' model does not work for persistent bioaccumulative chemicals because Nature quite simply collects and concentrates these materials over time.

What is needed is a shift from 'permissive' regulations based on attempted control of exposure and risk, to one based on prevention. The goal of chemicals policy should be the elimination of exposure to intentionally manufactured substances whose intrinsic properties give cause for high concern.

European chemicals regulations are currently being completely overhauled and new legislation in 2004. But the EU's proposed new chemical policy does not as yet include measures that will move us away from a permissive regime. Although the framework (REACH) and mechanism (Authorisation) are there, as it stands, the draft legislation continues with 'adequate control' as the regulatory paradigm. Continuing human exposure to may cause cancer, or genetic damage, endocrine disrupting chemicals and substances that build up in our bodies, is set to be tolerated.

We know there is a problem, what is the solution? The most important step towards a preventive regime, one that truly has protection of human health and the environment at its core, is to give a central place in chemicals legislation to the Substitution Principle. This can be defined quite simply as 'the substitution of hazardous substances by less hazardous substances or preferably non-hazardous substances where such alternatives are available'. It means that if a product that uses a hazardous chemical can be manufactured using a safer alternative, at a reasonable cost, the hazardous substance will no longer be permitted for that use. Common sense? Yes, but currently things do not work that way, and many hazardous substances are used without need, simply because there is no legislative or economic reason for substitution to take place systematically. Is the Substitution Principle workable?

Some companies are already using substitution as a means of eliminating hazardous chemicals from their businesses A variety of reasons exist for why some companies are searching for safer substitutes and these include regulatory drivers such as the recent Directive on the Restriction of Hazardous Substances, increased public awareness, demands from downstream users or clients, liability issues, competitive advantage and company ethics. However, there are also barriers and the development and adoption of safer substitutes is happening only slowly, in a piecemeal fashion and in some sectors not at all.

For this reason, the Substitution Principle cannot be implemented simply as a general policy statement, since this will be an insufficient driver for change. Instead it needs a clear mandatory imperative to drive it. Within reach this means that the substitution principle needs to be written into the authorisation procedure so that the availability of a safer alternative is sufficient grounds for an authorisation to be refused.

A safer alternative is one that does not meet the requirements for a chemical of very high concern as defined in Point 44 (a) to (f) of the REACH consultation document. 'Availability' means the substitute must be available on the market and is defined to include an economic element (i.e. at a reasonable cost). It must also be technically effective and fit for the use to which the application applies.

An alternative may be safer in that it does not meet the criteria for very high concern, but it may have other hazards such as corrosivity or flammability. These hazards are easier to control, but if there is a serious health and safety issue with a proposed substitute, that alternative would not be deemed an available alternative. A proposal for a workable authorisation procedure based on the Substitution Principle is shown schematically in figure. When an application for an authorisation is made, the applicant should provide details of alternative substances, materials, processes or products currently in use. A comparative hazard assessment of alternatives should be provided. Other parties (e.g. manufacturers of potential substitutes) should be invited to respond to this Substitution Assessment.

If the manufacturer, importer or user of a chemical of high concern can demonstrate that no viable alternative is available, that there is need for the chemical (with a transparent socio/economic assessment) and that the substance can be adequately controlled, a time-limited authorisation may then be granted. A time-limit will both ease costs of a phaseout and encourage development of alternatives.

The basis of this system is the presumption that a chemical of very high concern will be phased out unless the applicant can demonstrate that there is no safer alternative, there is a compelling reason for production to continue and that the risks can be controlled. Only in those circumstances would a timelimitedauthorisation be granted.

The availability of alternatives should not be seen as one consideration in the authorisation process, it should be the decisive factor in deciding the future of a chemical of high concern. Producers and users of chemicals of very high concern should be required to assess alternatives to these chemicals, and where none exist to develop them. This requirement will redirect the focus of chemical producers and downstream users towards safer chemicals and processes and stimulate the development of Green Chemistry. It will be a major driver towards Clean Production and effectively level the playing field for companies who are already implementing safer materials in their products. Without such a regulatory driver, however, the adoption of safer substances will proceed in a piecemeal fashion and entrench others in extensive risk assessment to justify ongoing use of chemicals of very high concern. Indeed the requirement to assess substitutes simply changes the focus from trying to manage and justify intrinsically hazardous chemicals to one that advances the use of safer processes and products in our society.

Background

Toxic contamination We are all contaminated by a wide range of chemicals. Each generation is progressively exposed to persistent chemicals in the womb and intake of persistent organic pollutants by nursing infants is high. Effects caused during development may cause permanent irreversible damage and some effects may not even become apparent until later in life. This is no longer acceptable, particularly when safer product design and chemical substitution is feasible.

Chemicals such as brominated flame retardants have become widespread contaminants and are now detectable even in marine mammals from remote areas as well as more generally in human blood and breast milk. Release to the environment occurs not only during the manufacture of these chemicals but also during the routine use and final disposal of a wide range of household and consumer products such as computer components, sofas, carpets, and textiles. Studies show the flame retardants, PBDEs, to have similar effects to PCBs by affecting neurological development in the foetus. (Allsopp et al, 1999) Again, the youngest in our society are generally the most susceptible to such contamination.

What is the Substitution Principle?

The Principle of Substitution states that hazardous chemicals should be systematically substituted by less hazardous alternatives or preferably alternatives for which no hazards can be identified. As a basis for protection of human health and the environment from chemicals of concern, the Substitution Principle has several advantages:

- It provides a stimulus and direction for innovation. Governments need not prescribe particular alternatives, but rather they define criteria to guide the identification of alternatives. This will end the common practice of replacing a hazardous chemical under legislative pressure with a similar hazardous chemical that is less in the spotlight.
- It implements the Precautionary Principle. Arguments against the Precautionary Principle commonly centre on what degree of evidence of harm is necessary before action is taken to restrict the use of a substance.

  When applying the Substitution Principle, it is not necessary to wait for cancers, reproductive disorders or genetic defects and elusive evidence of cause and effect. If alternatives with less hazardous intrinsic properties are available, use of the hazardous substance is not permitted. The potential for harm is then reduced or avoided altogether.

- It will avoid the current 'paralysis-byanalysis' syndrome where we have upwards of 30 000 chemicals on the market that have not been adequately tested. Substitution reduces the need for cumbersome, time-consuming, subjective and datapoor assessments of risk. If an alternative is available which is intrinsically less hazardous, intensive assessment of the original hazardous chemical is unnecessary.
- In many cases hazard based substitution eliminates the need for notoriously difficult exposure assessment. Persistence and bioaccumulative potential are surrogate measures of both hazard and exposure that can be applied quickly to all chemicals.
- The Substitution Principle provides the stimulus for Clean Production and sustainable product and system design. A focus on substitution also opens up other possibilities for solving the dilemma of how to replace chemicals of high concern. For example, where necessary the substitute may not be another chemical, but the redesign of a product. Computer companies are adopting metal framed housings to avoid the use of toxic flame retardants in plastic. Fire services point to preventative sprinklers, product redesign and alternative materials to achieve fire prevention in buildings. The same function has been achieved with safer alternatives.

Why not control the risks instead of insisting on substitution?

The considerable inherent uncertainty in our understanding of the way that chemicals interact with the environment means that there will continue to be a risk of serious effects, as a result of the use of chemical products, that we cannot predict on the basis of our current or foreseeable understanding of these processes. This requires a precautionary approach to chemicals management, and this is best implemented through substitution.

"We recommend that the UK Government adopt substitution as a central objective of chemicals policy." UK Royal Commission on Environmental.

It is not possible to adequately control the risks of persistent, bioaccumulative chemicals. These intrinsic properties mean that there is a high risk of exposure at sometime during the lifecycle of the chemical or the article that contains it. Even small releases, if they are continuous, can result in significant exposures. This is why we see escalating levels of brominated flame retardants, nonylphenols and other persistent chemicals in breast milk, umbilical cord blood and other human tissue.

Substitution, rather than risk management is therefore essential. Chemicals identified as of high concern, e.g. carcinogens, reproductive toxins, those that persist and bioaccumulate n the environment and affect the hormone system, should be targeted for substitution ased on their intrinsic hazards. ' It has been suggested by producers of rominated flame retardants that the health risks related to the extrusion of plastics containing PBB and PBDE could be avoided by strengthened worker protection measures in the recycling installations. As an example it was recommended that workers carry protection asks. Clearly, the substitution of the concerned substances would provide the best protection of the concerned workers. European Parliament and Council, 2002. By defining chemicals of very high concern, reach has already clarified substances that need to be phased out. Any chemical requiring an authorisation should immediately be targeted for substitution. This is a natural progression from what some member states have already suggested. In August 1999 a government decision in the German Bundestag called for a staged but drastic reduction in discharges of endocrine disrupters drawing on a similar decision by the European parliament on 26th January 1999. The UBA stated: Substances whose endocrine potential has been shown in in vivo tests, but where the available data is (as yet) insufficient for legal restriction or prohibition, should be named publicly in blacklists, and made subject to a substitution requirement under the Hazardous Substances Ordinance. Such a list could provide sufficient incentive to substitute, even where there is only a suspicion of danger. The Swedish government has also reiterated the need to substitute substances of very high concern. These substances are estimated to be relatively few and would not overload the authorisation system. Sweden then states that authorisation of such chemicals may be granted only if industry can demonstrate that no feasible alternatives are available; the socio-economic arguments clearly outweigh the potential risks and emissions are foreseen to be negligible during manufacture, use and disposa.

What is a 'safer alternative'?

The goal of substitution is to progressively move to safer materials and system design, so it is important to have a clear decision process and set of criteria to define what is 'safer'. As information has increased over the last few decades, there is a general tendency to move from halogenated chemicals to non-halogens specifically because many organohalogens are toxic, persistent and bioaccumulate in living systems or give rise to by-products with these properties throughout their different product life cycles. As more information emerges on endocrine disruption and neurotoxicity, suspect chemicals are highlighted. This has resulted in lists of restricted chemicals drawn up by both regulatory and industrial bodies. For example the OSPAR list of priority chemicals has focused much attention on the search for safer substitutes for all uses of these chemicals. However reach moves the list approach forward by setting a clear standard across Europe on what must be substituted. Any chemical meeting the criteria of 'very high concern' because of its inherent hazards should be subject to the substitution procedure. of 'very high concern' is a candidate for a safer alternative. There will undoubtedly be cases which are not as simple as this. For example, an alternative may be safer in that it does not meet the criteria for very high concern, but it may have other hazards such as corrosivity, flammability or volatility. However, these hazards are generally easier to control and fall under appropriate health and safety regulations. If there is a serious health and safety issue with a proposed substitute, that alternative would not be deemed an available alternative.

Substitution can be performed on several levels. The replacement of a hazardous chemicals by safer substitutes is usually accomplished by one of three procedures. Chemical (or material) substitution is the direct replacement of one substance for a hazardous substance, in a simple 'drop in' process. Many retailers, for example, have replaced PVC plastic with non halogen polymers or natural materials. Process substitution achieves the same results with less hazardous materials, such as the replacement of chlorinated solvents with aqueous cleaning techniques.

Function substitution achieves the desired outcome with a change in function or system. The use of metal casings in computers as a heat sink and barrier to flammable components is a functional substitute for plastics impregnated with brominated chemical flame retardants since they both achieve the same goal (fire prevention) but with different functions. Similarly a change to organic agricultural techniques such as companion planting achieves the same free crop conditions as from initial use of toxic pesticides.

Assessment of Alternatives

Risk assessment is the conventional tool for decision making on the acceptability of chemical use. It is based on predicted exposure levels, predicted no-effect levels of individual chemicals and politically defined degrees of acceptable risk. Each of these processes involves a series of value judgements and estimations. Risk assessment is therefore highly subjective.

On the other hand, the assessment of intrinsic hazard is a much more objective process. Intrinsic properties can be established and quantified and provide a better basis for decision making.

The assessment of substitutes will rely on various factors and a number of methodologies have been developed to compare the alternatives available. A variety of software and tools have been developed to enable a comprehensive review of chemical properties. One model, developed for the German Ministry of Environment, uses an evaluation matrix based on indicators to compare a range of criteria, such as specific chemical properties, intrinsic mobility, amount used and indirect releases into the environment. Weightings are given for each chemical assessed and then results are placed on a risk index ranging from very high to very low.

The evaluation matrix then ranks the most preferable. The indicators for the evaluation and a software tool for carrying out an evaluation can be downloaded from the internet Additionally, the German government has studied in some detail the drivers and barriers to sustainable substitution of cals. Another model, used in the USA, is the OASIS Pollution Prevention Optional Analysis System, developed by the Toxic Use Reduction Institute in the USA. Here, technologies are assessed on a variety of hazard criteria, including acute and chronic human, physical, aquatic, persistence/bioaccumulation, atmospheric, disposal, chemical, energy/resource use, product hazard and exposure potential. Alternatives are rated to display a series of scores and the final weighted score displays the best option for the manufacturer. The Swedish Chemical Inspectorate (KEMI) has used substitution analysis to assess biocides with the goal of promoting sustainable agriculture. KEMI uses a seven-step process which evaluates not only the intrinsic hazards of a chemical but its efficiency, its cost and its intended use. Alternatives are monitored and assessed for effectiveness. Since the Substitution Principle has been operational, 20% of the pesticides on the Swedish market have been substituted with less hazardous products. Users have a reasonable time frame to adopt the substitute and this enables smaller pesticide formulators and farmers to adapt to the change with few complications. When a safer product is put on the market KEMI allows the existing and more hazardous product to be used until its approval expires many progressive companies have used the substitution Principle to move towards the goal of clean production and some downstream users of chemicals claim to only use benign chemicals in their processes and products. For instance the McDonough Braungart team of consultants have worked with the Design Tex company to produce a carpet which is made with non hazardous chemicals from their 'Positive' chemical list Other examples from the European Roundtable on Cleaner Production have been gathered and examples of successes in Green Chemistry, as listed in Annex I, demonstrate the feasibility of designing benign products.

Few people would argue that substitution is undesirable but without a legal driver it happens only in a piecemeal fashion. The objective, to eliminate all but absolutely necessary uses of chemicals of very high concern is not achieved; The Substitution Principle must be an obligation for all not only for the most progressive companies Substitution is already a goal for some progressive companies. It helps them to systematically reduce their use of hazardous chemicals and develop new products. Some companies have agreed on substances that need to be avoided. As documented in the case studies in Annex I of this report, they are instructing their suppliers to phase out a range of carcinogenic, mutagenic and reproductive toxins as well as some persistent bioaccumulative substances and endocrine disrupting chemicals. However, voluntary substitution will not end general exposure to chemicals of high concern. Several studies have shown that voluntary actions have severe limitations.Recently, the OECD published a report pointing out some of the major problems with voluntary as opposed to obligatory schemes. In particular they point out that 'free riding' is a significant problem among many collective voluntary approaches. This fact was highlighted during industry's lobby for the Directive on the Restriction of Hazardous Substances and the industry sector was adamant that restrictions must apply equally across Europe.

"We urge the Council to support the current legal basis of the Commission's Waste Electrical and Electronic Equipment and RoHS directives. The European Parliament has confirmed that Article 95 should remain as the legal basis for the RoHS directive, so as to provide a high level of protection for citizens without creating uncertainty for business and undermining the Single Market. Environmental protection and consumer safety are equally precious for all EU citizens. An improvement affecting all 350 million citizens is more desirable than only a few countries going ahead with divergent national legislation." Joint letter by Electrolux, HP, ICL, Nokia, Ericsson, Agilent, IBM, Gillette, Sun Microsystems, and Intel. to the European Commission The OECD further note that the economic efficiency of voluntary approaches are generally low because they fail to equalise the costs between all producers and environmental targets are set on individual sectors rather than at a national level. While the proactive efforts of some industry demonstrate that substitution is indeed feasible, voluntary action will not solve the global problem of increasing chemical contamination. Such action needs to be universal and across all industry sectors and size of firms. Industry needs clear criteria in which to operate and innovate. In particular, small and medium size enterprises need clear criteria to chart their way forward in a competitive economy. Substitution regulation spurs innovation.

Just as sustainability presents us with the most troubling and complex technical challenges we face, it also highlights the most important technological opportunities crying out to be cracked by today's chemists. Finding the solutions will result in major economic progress.

Terry Collins, Director of the Institute for Green Oxidation Chemistry, The role of regulations in promoting innovation has been a matter of discussion for a long time. The Royal Commission on Environmental Pollution in the UK appointed the Science and Technology Research Unit at the University of Sussex to report on the impact of regulations on innovation in the chemical industry. The report concluded that the introduction of new regulation sometimes porary decrease in innovation activity, but in the long term has no negative influence on innovation. It notes that in many countries the most successful firms and industries are those that face the highest levels of regulation. The 1982 Swedish ban on the use of cadmium as a pigment, surface treatment and stabiliser, backs up this conclusion. The ban was implemented in the face of strong protest from companies claiming the costs would be too high. Later, an assessment of the effects of the ban was undertaken. It was discovered that the ban had caused some short term financial difficulties for some companies, but in the longer term, no effect was seen on the companies' market shares or profits. In fact, regulatory drivers are often seen as instrumental in spurring innovation. An assessment by a major aerospace manufacturer of what drives innovation towards safer processes for hazardous waste streams revealed that: 'The regulatory drivers for waste reduction are familiar to most by now, and may be summarised into three categories of legislation:

1. inventory reporting;
2. emission reporting; and
3. employee exposure levels. Anticipation of future restrictions was a decisive factor in this project. The mandate within the recently passed.

Directive on the Restriction of Hazardous Substances (RoHS) to phase out a range of hazardous materials in new electrical and electronic products by 2006 has been the most significant driver to product redesign not just in Europe but in other regions also. For example, an assessment of the high tech industry in Asia has revealed that impending European legislation was the driver for hazardous chemical phase out and adoption of alternative designs and chemicals. While manufacturers in Europe were opposing the impending WEEE and RoHS directives, Japanese companies strategically positioned themselves as promoters of greener products, and took the lead in finding substitutes for lead solder in electronics.

As a result Japan developed lead-free soldering technologies well ahead of the WEEE directive timescale, and ahead of European and American counterparts Effects of the substitution requirement on pesticide registrations in Sweden Since 1986, the Swedish Chemical Inspectorate has required the registration of pesticides. Beginning in 1990 a re-registration every 5 years for pesticides included a comparative assessment of substitutes. The year before the re-registration started, 619 pesticides were registered. The introduction of the substitution requirement caused a temporary decrease in numbers and only 343 appeared on the market the year after. Some pesticides were rejected for re-registration, others were substituted with better alternatives and for others, the producers did not apply for re-registration because they realised that their products did not fulfil the requirements and would be substituted. The temporary ecrease in pesticide use lasted a few years, but today more than 700 pesticides are registered, demonstrating a wider range of products which are safer for human health and the environment.

Regulatory drivers are needed to stimulate the research and market for safer substitutes. If the cost of a safer substitute is too high due to lack of market demand and interest mong companies within a sector, the innovation will remain marginal at best. In particular, it may not spread to the SMEs who would face an even more difficult financial challenge to adopt the safer substitute. A clear regulatory focus on substitution would stimulate the research, development and adoption of safer processes and products. For example, it is well documented that impending legislation under the Montreal Protocol to ban ozone depleting CFCs stimulated the widespread research and adoption of non ozone depleting alternatives.

Similarly, the various EU directives that mandate substance bans have initiated wide scale research and development of alternative materials. Lead is a case in point. Lead has been widely used in the electronic industry for use in solders. Lead-free solders have existed for many years but it was the mandate in the Restriction of use of certain Hazardous Substances Directive to have products free of lead by July 2006 that spurred industry research, planning and adoption of the substitutes examples of Green Chemistry case studies where research was stimulated in response to legislation targeting hazardous materials. Legislation is needed to ensure that data is available for substitution purposes.

A common complaint from downstream users of chemicals is that data gaps prevent adequate characterisation of the risks from chemicals. Increased information under reach will change this situation as long as data are transparent, readily available and accessible to the public. But collecting information on chemicals in the absence of a clear goal to replace chemicals of concern is likely to lead to the entrenchment of hazardous chemical use within a more costly end of pipe management control. Any system that does not forc e an examination of safer substitutes and make this a priority over the risk management of hazardous substances, will ultimately fail to protect people from needless exposure to chronically dangerous chemicals.

Perchloroethylene in dry cleaners Perchloroethylene is a probable human carcinogen and neurological toxin which poses risks to both workers and consumers, as well as a ubiquitous groundwater contaminant. A variety of safer substitutes have existed for over a decade which the Canadian and US governments have independently monitored. Both studies confirmed the economic and environmental benefits and technical feasibility of these alternatives. However, the majority of dry cleaners continue to use perchloroethylene ecause there is little awareness and no legislative imperative to choose safer substitutes.

Researchers on substitution for the German government have documented a variety of reasons why substitution fails to be carried out. For example, they detail how cement manufacturers in Scandinavia solved the problem of skin contact with cement containing hexavalent chromium back in the 1980s, but the information was not diffused to other European manufacturers. Legal precedents for the Substitution Principle The acceptance of the Substitution Principle as a workable legal act was demonstrated in a European Court of Justice (ECJ) court case in 2000. Trichloroethylene (TRI), a cancercausing chemical, was banned in Sweden and companies had to find alternatives.

Exemptions were only given when a suitable alternative was not available, when use did not lead to unacceptable exposure and on the condition that the company continued to seek alternatives. For the majority of exemptions, the firms had managed to substitute TRI in most of their production, but had not found a suitable alternative for a specific use in the production process. One firm appealed against the ban, but the European Court of Justice ruled against them. The ECJ ruling demonstrates acceptance of the Substitution Principle in EU courts.

Mandatory Planning for Substitution

Making companies prepare plans which focus on safer chemical use has proved particularly successful in the USA.

The benefits of mandatory pollution prevention planning have been demonstrated in the state of Massachusetts. Here, over 550 companies had to assess toxic use reduction options with technical help supplied by a university and government experts. Toxic use reduction strategies included material substitution and product reformulation. Within ten years, industry has reduced the use of toxic chemicals by 40%, by-product waste by 58% and toxic emissions by 80%. A cost benefit analysis reveals that the same companies saved a total of 14 million dollars over this period through the adoption of more efficient and safer processes. The program is ongoing and has been expanded to community outreach and assessment of substitutes for some hazardous material flows and products within the state. When a company applies for an authorisation it should be required to provide an assessment of available substitutes. This should include a full description of the alternatives available, a comparative assessment of their intrinsic hazards and an assessment of technical feasibility. This substitution assessment would form the basis of a justification of why an intrinsically less hazardous alternative cannot be used to replace the substance of high concern. The Substitution Assessment should be transparent as to the methods and data used in seeking and assessing alternatives.

If the applicant demonstrates that no suitable alternative is available, and can satisfy the other requirements for an authorisation to be granted (a social need, a positive cost benefit analysis and adequate control) a Substitution Development Plan should be required, so that chemical, process or function substitution can take place at upon expiry of the authorisation period. The Substitution Assessments for each individual use of an authorised chemical should include a range of alternatives based on chemical, process or function substitution. Cost factors may initially be higher than continued use of a chemical of high concern, but increased demand for the alternative will drive costs down, particularly as competition increases among producers to supply the new market demand. Chemical producers in turn will find an expanded market for Green Chemistry products.

To further stimulate the drive to safer substitutes, a fee could be levied on users of all authorised chemicals. This fee could be used to centralise and disseminate information on alternatives, and fund research. Expertise to help companies already exists in many countries. Companies who are adopting safer alternatives often contract outside help.

Chemical producers have their own in-house research teams. Other institutes work with SMEs such as the European Cleaner Production networks and some member states have well established programmes which focus on sustainable product design and safer chemical use. Some member states give prominence to substitution in government policy. For example, the Danish environmental strategy prioritises action on their dangerous substances list and encourages manufacturers and importers to find substitutes and to develop alternative products. The Danish EPA's 'Cleaner Products Support Programme' grants subsidies to a number of projects that promote substitution. It supports the development, testing and assessment of alternatives to brominated flame retardants, as well as the dissemination of knowledge to manufacturers about the feasibility of implementing alternatives.

Finally it is essential that ALL authorisations are time-limited. This is necessary to move towards the goal of phasing out all chemicals of very high concern. Annex I to this report exhibits a large number of cases, where substitution has been carried out successfully, following a systematic approach to finding alternatives. The information was compiled from paper and on-line research and from conversations with industry representatives. Other case studies show how leading retailers are moving to phase out hazardous materials and how some chemical companies are researching and implementing Green Chemistry. The data demonstrate that substitution is feasible and is already happening in the more progressive sectors of industry. For example: Hewlett-Packard monitor housings now typically contain phosphorous-based flame retardants, and its computer casings have no brominated flame retardants. Bayer, a major supplier to the electronics sector, supplies an ABS/polycarbonate blend which contains flame retardants based on triphenyl phosphate or phosphate derivatives instead of bromine compounds. Some manufacturers use internal metal 'shields' to protect computer housings from internal sparks and heat, and can therefore use non flame retarded plastics. The most recent example of this is Apple. The next generation of Apple products will move away from polycarbonate housing and towards metal housing, using an aluminium alloy Until recently, halogen-free products were only available in Europe, but Sony has now adopted global design standards to ensure that all their projects meet the same standards.

Sony aims to have all product lines free of brominated flame retardants by the end of 2005 if substitutes are found to be safer. They also aim to phase out all uses of vinyl chloride by 2005 as well as lead solder, and specified heavy metals. For example, a Walkman model has PVC free cables and no brominated flame retardants or lead solder in the printed wiring board.

Marks and Spencer initially targeted the successful phase out of pesticides in their product lines and then turned their attention to PVC. By eliminating PVC, they effectively eliminated most phthalates. Their transition from PVC in handbags, belts and shoes led to significant cost increases, but they have achieved a full phase out as well as an 89% phase out of PVC in packaging. Alternative plastics such as PET were used in some instances. Boots decided to phase out alkylphenols in 1999 and by 2000 they had achieved 90% phase out.

Other than a phase out of all brominated lame retardants, H&M (Hennes and Mauritz) have achieved a strict phase out of alkyl phenol ethoxylates, organotins, azo dyes and all carcinogenic dyes, PVC, bisphenol-A, phthalates, antimony and a wide ranges of heavy metals as well as chlorinated aromatic hydrocarbons.

Companies with experience from working cases actively promoting the inclusion of the principle in legislation. This is for instance seen in comments from companies to the EC July 2003. Notably, companies with lengthy experience from national substitution requirements, praise the legislative instrument, and promote its use in reach. Skanska, one of the world's largest construction companies with 75 000 employees and activities world-wide, are stating that operating for many years under substitution regulation in Sweden has lead them to '...continuously seek less harmful alternatives. This is something that our clients expect from us as a producer of buildings or infrastructure. As we are not experts on the components in our products, we have to go back to our suppliers with the requests that our clients put on us. As manufacturers of building components they will have to go back to their suppliers etc. This is the way we want the market to work in order to reduce the environmental impact'. Skanska are disappointed with the lack of a strong substitution rule in reach: 'The present proposal is much more static, focusing only on evaluation and registration. Without the strong support from a Substitution Principle it will be difficult for an individual company that is a downstream user to be proactive in substituting substances."

Tetra Pak, the world's largest producer of food packaging, take a similar stand: 'These two principles are important principles in the Swedish national chemical policy and have proven to be a good basis for chemical control. Tetra Pak is therefore supportive to building the REACH system on these two fundamental principles. Precaution and substitution need to be introduced early in the text as guiding principles for the whole policy.'

In fact, the collective construction industry in Sweden are strong advocates for an EU-wide substitution regulation. The Construction Federation which represents the interests of the construction industry in Sweden, state in their response to the reach consultation that: 'Particularly hazardous substances must be blacklisted and, accordingly, banned. Only substances for which there is no safer alternative may be exempted and this only if there are strong social or economic reasonsfor doing so. If an exemption is granted, producers/users must take precautions in order to minimise the risks.' They emphasise that: 'Products containing particularly hazardous substances should not be authorised just because the producer/importer can show 'adequate control'. Substitutions should always be considered." The Federation calls for the following text on the Substitution Principle to be incorporated in the law text: 'Particularly hazardous substances must be blacklisted and banned.

Exemptions may be granted only when it is clear, after a thorough investigation, that there is no safer alternative and if the social and economic advantages are greater than the risks involved with using the substance.

Industry sectors further down the product chain express their concern about a weak or non-existing Substitution Principle in reach. The Recycling Industry Association in Sweden, for instance, considers both precaution and substitution practically absent in the legislative proposal. They insist that 'In order for EU to maintain a high level of chemical control, it ought to be explicitly stated in introducing articles of the legislative ext, that the Precautionary Principle and the Substitution Principle constitute the fundamental principles of the entire legislation.

The principles also ought to be incorporated into the Duty of Care chapter, to make clear that all parties will follow the principles, and be responsible for having sufficient knowledge to uphold the duties set up by REACH. H & M, one of Europe's largest retail chains, is a strong proponent of safer substitutes.

They state: H & M is applying the precautionary principle. In practice, this has meant working closely with our suppliers to phase out substances and materials, that are, or could potentially be, harmful to our customers or the environment, from our products. In doing so, we have constantly, together with our suppliers, searched for less harmful solutions.

We have encouraged our suppliers to be innovative and when we have found a better alternative somewhere among our suppliers we have helped to spread that knowledge to other suppliers and other markets.

In doing so, we have found that almost anything is possible as long as you set clear guidelines on what is not acceptable. We have not had to compromise on fashion or quality in a way that has harmed our business. Prices may have gone up temporarily but as soon as mass production has started, the prices have gone back to previous levels. With the background of this experience, we find it important that EU legislation supports the idea of substitution when a better alternative is available. Such legislation would support us in our continued effort to eliminate hazardous substances from our products and to find better solutions that are less harmful to the environment.

Ingrid Schullström H & M ASE STUDY: Electrolux Electrolux is the world's largest producer of powered appliances for kitchen, cleaning and outdoors. They have created comprehensive.

Environmental Product Declarations (EPDs) for many of their product lines. Information within the product profiles details chemicals which have been banned as well as the percentage and types of materials and how they have improved material choices. For example, plastics do not contain cadmium, lead, mercury or their compounds or chlorinated or brominated flame retardants; metals are not coated with cadmium, chromium, or nickel; and metal paints do not contain pigments and additives based on heavy metals. Many Electrolux products are also PVC-free.

The proposed REACH legislation attempts to address the lack of information on existing chemicals and the need to prioritise chemicals of high concern for regulation and control. However, under the current draft, even the most hazardous chemicals will be authorised for continued use if a manufacturer can demonstrate that the risk to human health and/or the environment is 'adequately controlled'. If adequate control cannot be demonstrated an authorisation may still be granted if socio-economic benefits outweigh the risk to human health and/or the environment arising from the use of the substance. This decision shall be taken after consideration of: the risk posed by the uses of the substance; the socio-economic benefits as demonstrated by the applicant or other interested parties; and any available information on alternative substances or technologies. Legislation in this form will mean that even the most hazardous substances will be granted authorisations for continued production, even when safer alternatives are available. 'Adequate control' does not prevent releases into the environment and the intrinsic properties of chemicals of very high concern mean these releases will lead to ongoing exposure and continued build up in the environment and in human beings. If reach allows the continued production of chemicals of very high concern under a provision for 'adequate control', even when intrinsically less hazardous substitutes are available, it will not provide the high level of protection for human health and the environment required under the EU Treaty.

The Substitution Principle should be the key principle of the authorisation process. Specifically, it must be mandated that the availability of a safer alternative is in itself sufficient reason to refuse an authorisation. This is the only way to ensure reach is a safer chemical production and innovation; not an entrenchment of hazardous chemical use that permits the continued, unnecessary exposure of people to hazardous chemicals for the sake of short term profits. Some sectors of industry are already developing practical programs on substitution. At the same time, expertise exists within the European Union to help small and medium sized companies implement safer products and processes.

A requirement to provide a Substitution ssessment with all applications for an horisation will prevent unnecessary requests for authorisation and focus attention of safer chemicals. If substitution is not currently feasible for a particular use, the use of an authorised chemical would be allowed under a strict risk management regime, if social need could be demonstrated and a positive cost/benefit analysis provided. The authorisation would be time limited to allow the development of safer substitutes, and manufacturers and/ or users would be required to produce a substitution development plan to enable substitution to take place before the authorisation expires. Such planning for chemicals of very high concern would vastly increase the information flow and development of safer substitutes. It would also move Europe to a more competitive, innovative and sustainable producer of chemicals, goods and services.

More importantly, it would begin to reverse urdens of hazardous chemicals that we all now carry. The following case studies focus on chemicals likely to be defined as of 'very high concern'; i.e. they would require authorisation under the REACH system.

The information was compiled from paper and on-line research and from direct phone conversations with industry representatives he more comprehensive overview of case studies of company substitutions for brominated lame retardants demonstrates both the complexity and feasibility of implementing safer alternatives. Other case studies show how leading retailers are moving to phase out hazardous materials and how some chemical industries are researching and implementing Green Chemistry alternatives.

Substituting brominated flame retardants in products Brominated flame retardants (BFRs) are used in a wide range of consumer products: electronic components, textiles, foam in upholstery and carpets, and building materials – all uses where the risk of fire necessitates caution. The increase in the use of more plastics and flammable synthetic materials has contributed to the increase in the use of flame retardants.

As evidence grew by the late 1980s of the dangers of BFRs, particularly PBB and PBDEs, Germany, Denmark, the Netherlands and Sweden began restricting and banning their use. In a declaration of intent in 1989, the chemicals industry and plastic manufacturers in Germany declared that they would neither produce nor use PBDEs.

The electronics industry found alternatives to BFRs The electronics industry moved quickly to find alternatives ranging from material substitution (replacement of halogenated flame retardants with non-halogens) to functional substitution (replacement of plastic with metal housings). Much of the stimulus for better design and less hazardous materials has come from the Waste Electrical and Electronic Equipment (WEEE) and Restriction of the Use of Hazardous Substances (RoHS) directives and their emphasis on recycling and chemical bans. 80% of a typical product's environmental impact is determined by its design. As concern around bromine compounds grew, industry moved away from those under the greatest legislative pressure (PBDEs and PBBs). However, as the market for PBDEs and PBBs has declined, sales of bromine compounds such as TBBP-A and HBCD have gown in equal measure. Meanwhile, understanding of the toxicity and persistence of TBBP-A and HBCD has increased the pressure to address BFRs as a class, and in parallel the development and supply of nonhalogenated chemicals has increased. As information grew on the risks and feasibility of some of the phosphorous based alternatives, some companies developed nonphosphate alternatives. Product design is increasingly looking at nonchemical alternatives to flame prevention, such as the use of metal casings instead of plastics. Hewlett-Packard monitor housings now typically contain phosphorus-based flame retardants, and its computer casings have no flame retardants at all.

Bayer, a major supplier to the electronics sector, supplies an ABS/polycarbonate blend which contains flame retardants based on triphenyl phosphate or phosphate derivatives. Some manufacturers use internal metal 'shields' to protect computer housings from internal sparks, and can therefore use nonflame retardant plastics.

The most recent example of this is Apple. The next generation of Apple products will move away from polycarbonate housing and towards metal housing, using an aluminium alloy. Life cycle assessments show that although mining of metals increases energy use, the use of metals provides for better endof- life management (than plastics), provides a better heat sink (than plastics) and enables the product life to be extended. The use of aluminium alloy shows little weight change, compared with polycarbonates. In addition, the use of lacquers in metal finishes does not cause problems during smelting, but the use of lacquers on plastics renders the plastic non-recyclable. The shift from plastics to metal alloys was in part driven by the flame retardant restrictions of WEEE and RoHS. Motorola is a multi-national company that produces hundreds of different electronic systems for a large range of industry sectors, including the IT and automotive sectors. As Motorola became aware of the inherent human health risks posed by the use of BFRs, the company worked with suppliers to find safer substitutes for their printed circuit boards. Despite the bromine industry claim that BFRs are necessary to achieve fire safety standards, Motorola is able to meet the standards while using safer materials. Most of their new products use a safer halogen-free flame retardant alternative that is a nitrogen/phosphorus combination. Motorola's research on the alternative flame retardants concluded:

- Eliminating BFRs from printed wire circuit boards (PWB) reduces risks associated with dioxin formation in fires.
- Eliminating BFRs enhances ISO 14000 performance.
- Some PWB resin systems with alternative flame retardants possess better electrical properties.
- Some PWB resin systems with alternative flame retardants possess better mechanical properties.
- Some halogen free PWBs have demonstrated compatibility with lead-free assembly.
- Cost curve is very acceptable and is projected to meet current costs for best in class FR-4 flame retardant standards.
- The availabilities of substitutes are acceptable and projected to get better.

In response to the German dioxin ordinance of 1994, Sony Europe started investigating safer substitutes for halogen-based flame retardants. Sony has developed halogen free circuit boards used in European television sets, VCRs and DVD players. Printed wiring boards use resin which is an inherently flammable material. Sony's engineers adopted a resin structure containing nitrogen to increase heat resistance and modified the content and dispersibility of the phosphate compounds and fillers. Since the circuit boards must be completely halogen free, Sony also substituted phthalocyanine green, which contains chlorine, with phthalocyanine blue as the photoresist pigment that covers the boards surface. By substituting all chlorine- and bromine-based chemicals with safer alternatives, there is no longer a risk of dioxin formation throughout the product's life cycle. Until recently, halogen-free products were only available in Europe, but Sony has now adopted global design standards to ensure that all their projects meet the same standards. Sony aims to have all product lines free of BFRs by the end of

2005 if substitutes are found to be safer. They also aim to phase out all uses of vinyl chloride by 2005 as well as lead solder, and the heavy metals listed in the RoHS Directive. Details of their phase out of hazardous materials is given for specific product lines in their end-of-year report. For example a Walkman model has PVC free cables, no BFRs or lead solder in the printed wiring board.

In 2001, Samsung Electronics Co. Ltd. developed a 'Green semiconductor' that uses no halogen compound or toxic substances such as lead, chlorine and bromine. The company was the first to develop a package and module that contains neither lead nor halogens. The alternative increased quality, and has saved 960 million won (684,000 euros) since its inception. Samsung has marketed its efforts in substitution to enhance its corporate image as an environmentally friendly company which responds rapidly to international environmental regulations. Information on alternatives is seen as confidential National/Panasonic (Matsushita) joined forces with other major manufacturers to develop electric wires and plastics that do not contain halogen compounds. In September of 1999, they began marketing the world's first widescreen television for which halogen compounds had been eliminated from low voltage internal wires, from the cabinet, from the back cover and from a number of printed circuit boards. At present, this technology is being successively applied to a wide range of other products, such as PCs and monitors.

The company does not state on its website what alternatives they are using. Matsushita was at the forefront in substituting lead in their semiconductors. In 1991 the company committed itself to remove lead by fiscal year 2002 and it has succeeded in doing so. The company's website gives details on alternatives used: mostly silver, tin and bismuth compounds. IBM's Design for Environment guidelines focus on 15 environmental attributes. These guide product designers of Personal Computing Division products.

The company mandates that resins and paints used on PC products must not contain polybrominated biphenyls (PBBs), polybrominated diphenyl ethers (PBDEs) lead, cadmium, or mercury metal. IBM gives its Original Equipment Manufacturer (OEM) suppliers, such as Samsung, a list of 11 banned PBDEs and PBBs but does not stipulate which flame retardants, if any, must be used and accepts what its suppliers manufacture. TBBA, a BFR which is reactive and thought to be bound to the polymer, is used in the epoxy resin laminate in printed circuit boards in most manufacturers' products. In 1997, a phosphorus-based alternative was developed by the German engineering giant, Siemens, n with support from the German Research and Technology Ministry. The laminate is manufactured under licence by Siemens-Nixdorf, a wholly-owned subsidiary of Siemens, and by the German chemical group Hoechst. Furthermore, the production waste, which can amount to 30% of the final product's weight, can be recovered more easily because of the absence of halogens. All of the housings and keyboards manufactured by Siemens-Nixdorf are BFR-free. The electronic giant, NEC produces mobile phones, office equipment and personal computers.

Its environmental policy includes a target to stop using halogenated flame retardants by 2011. In 1999 the company went one step further and launched a polycarbonate containing a silicone flame retardant which it claims to be 'far superior to conventional flame-retarding plastics in environmental safety' and is neither phosphorous nor halogen based. Sold under the brand name NuCycle, the new material is used to make NEC's liquid crystal display (LCD) monitors and battery packs for portable computers and it can be recycled up to five times for the same purpose. The material is manufactured under licence by Sumitomo Dow which supplies it to NEC at the same price as that of ordinary polycarbonate resin. Sumitomo Dow also supplies NEC with a polycarbonate resin which dispenses with the need for brominated or phosphorus-based flame retardants. Several Japanese firms – including NEC, Hitachi, Toshiba and Sony – have developed phosphorus-based flame retardants as alternatives to TBBA in circuit boards. Fujitsu- Siemens has eliminated TBBP-A from the maincard on their Scenic computers by substituting with phosphorous acid that is reacted within the laminate.

In 2000 NEC developed an epoxy resin with what it describes as a fire-retardant structure that avoids the need for either TBBP-A or phosphorus-based flame retardants in circuit boards. The new resin contains a metal hydroxide retardant. The company claims the new board is 'almost totally free of pollutants', and is easy to process and thermally recycle. By also integrating flame retardant properties within the board, use of the metal hydroxide is minimised, while offering good electrical properties, higher heat resistance and improved processing characteristics.

BFRs and the retail industry IKEA prohibits a range of hazardous materials in their product lines. They prohibit azo dyes in textiles and a product wide ban of BFRs and PVC. IKEA chooses textiles and materials that by nature are difficult to set on fire and can often completely avoid the need for chemical flame protection in their products such as interliners made of nonwoven inherently flame retardant materials.

When the company has to meet more stringent fire standards for the UK and California market, they employ chemical substitutes for some product lines. To meet UK fire standards which are the most stringent in Europe, IKEA replaced brominated compounds with organic phosphorous and nitrogen compounds either applied by impregnation of the cover fabric or applied to a cotton interliner. The work was done in conjunction with Ciba Geigy. Deca-BDE is used widely in polyurethane foam in the USA due to the lack of regulatory controls, and limited awareness of the dangers of BFRs within the upholstery industry. IKEA was able to meet the stringent Californian standards by switching three years ago to melamine combined with chlorinated paraffins. Melamine is not a bioaccumulative or persistent material. They are now researching alternatives to chlorinated paraffins, including the use of novel substances, such as expanded graphites and their suitability within foam. IKEA is conducting this research in isolation in the USA and points out that the cost of new substitution will be high unless other upholstery designers and retailers follow suit.

Marks & Spencer continue to develop substitutes for deca-BDE in some product lines. Their focus has been on PVC substitution, and that has now been completed. Suppliers to Laura Ashley have confirmed they do not use any BFRs in their product lines. H&M (Hennes & Mauritz) have chemical restrictions for textile, leather, plastic and metal products based on international regulations and some additional chemicals of concern. They test for compliance from all their suppliers in their own labs. H & M uses no flame retardants in any product line using instead natural materials that are inherently flame resistant. In a few cases the company cancelled clothing line items that proved a potential flammable hazard and could not be materially replaced. Leaders in the building trade phased out BFRs and focussed on functional alternatives Skanska, ranked by the Financial Times as the world's most respected company in the property and construction sector, was also one of nine construction businesses listed on the Forbes A-list. Skanska Denmark and Sweden have been leaders in the task of developing a national industry-wide chemical database. The Skanska database now contains 2, 416 products. On the basis of these, Skanska Denmark has then selected products that may be used in its own operations. The industry-wide database of chemical products is evolving all the time, and the number of chemical products listed is rapidly increasing. Skanska Denmark has further refined the standard version of the industry-wide database so that each individual chemical product is classified on the basis of a number of criteria, taking into account both environmental and working environment as well as quality and economic factors. Products are classified in three categories: permitted products; products that should be phased out; and non-permitted products. In total more than 7,500 chemical products have so far been evaluated.

Skanska is replacing BFRs with product redesign and material substitution. They initiated a project – 'Brominated Flame Retardants in the Building Industry' – to audit the use of BFRs in building materials, analyse substitutes, and draw up a plan of action. The results of their investigation found that fire standards for individual components should shift to the installation as a whole and be more function oriented. They identified several examples where purchasers selected a flame retarded product marketed as 'fire proof' even where no legal requirements supported the use of flame retardants. Assessing non-brominated flame retardant chemical alternatives – are they safer? The Bromine Industry in particular, points out that non-brominated chemical flame retardants may be very hazardous and that the phase out of BFRs may not be safe. The non-brominated alternatives listed in the above case studies are likely to be safer if they exhibit less persistence and bioaccumulation in living systems, but some do possess significant toxicity. In the last few years, some agencies have therefore investigated the relative merits of alternatives to BFRs.

The German Environmental Protection Agency (UBA) surveyed 13 flame retardants for toxicity to humans and the environment and their suitability for closed loop substance management. The aim was to assess the feasibility of substitution with less hazardous flame retardants. They selected red phosphorous, ammonium polyphosphate and aluminium trihydroxide as the least environmentally problematic alternatives Red phosphorus can technically be used in a variety of polymers to meet even the toughest fire safety standards, although it may not work for all applications.

UBA remarked that: 'It is encouraging that there is a general trend to refrain from the use of halogenated flame retardants in products and to replace them with less problematic lame retardants or to redesign flame retardant systems, e.g. by creating greater distances to potential heat sources.' The summary of their investigation lists flame retardants under the headings:

- Phase out is recommended;
- Reduction is expedient, substitution desirable;
- Problematic properties; reduction expedient;
- No recommendation possible due to gaps in knowledge;
- Use is unproblematic The Danish Environmental Protection.

Agency also investigated the profiles of 12 alternative flame retardants for environmental and health effects. From a survey of published literature, it was observed that the amount of data available is often very limited, particularly for important criteria such as degradation. Furthermore, the screening study showed that the majority of the alternatives also had undesirable environmental and health characteristics, but that an assessment was needed to determine the amount and the manner in which they are released before a conclusion could be drawn. EACH will obviously help to fill this much needed gap in information. It can be surmised that the phosphorous alternatives to BFRs listed in the case studies above are generally safer because of the non-organic, less bioaccumulative nature of the compounds. Some alternative compounds are in need of more research depending on the type of mix used. Specific information on the exact chemical composition was generally not available. However, it is vital to remember that, within the context of Substitution

Assessment planning, substitution is not a simple process. It also takes account implicitly of the need to develop effective alternatives where they are not already available and to adapt rapidly to technical progress.

Material and functional approaches to substituting BFRs Using alternative chemicals is only one route to safer substitution. Material and functional alternatives also exist as well as preventative action to ascertain the real need for flame retardants. An analysis of possible substitution choices for BFRs in the computer and auto industry was conducted in Germany and subsequent stakeholder sessions were held to further discuss the types and feasibility of alternatives Substitution fell into the following basic types:

- *Using non-flammable materials:* Merely substituting flammable with non-flammable materials, e.g. plastic with ceramic circuit boards, can render the use of flame retardants unnecessary.
- *Preventing fire risk by improving design:* Increasing the distances between possible flashpoints and flammable materials may be sufficient.
- *Scrutinising fire safety regulations:* Are existing fire safety regulations genuinely justified, when the risk of fire is weighed gainst the risks associated with flame retardant chemicals (e.g. keyboards)?
- Substituting hazardous flame retardants with products that have less impact on the environment and human health. Some companies have gone down the nonchemical solution route: e.g. IKEA and Sony's use of non-flammable materials. The issue of fire safety regulations is topical. The bromine industry maintains fire safety is paramount and particularly defends the increased rate of some of its chemicals in the USA by pointing out that less stringent fire standards in Europe are associated with more fire deaths.

An examination of fire death rates for countries in Europe and elsewhere reveal that in fact deaths by fire are no higher in Europe and that the USA has higher number of fire deaths per 100,000 persons than many European countries Europe, on the whole, uses a fire standard based on a smouldering cigarette test, whereas California uses criteria based on direct flame tests.

The disparity of opinions about European and American approaches to fire safety and standards lead to the creation of Green Flame, a program of the International Consortium for Fire Safety, Health and the Environment. The Swedish Rescue Services Agency, the Swedish National Chemical Inspectorate, the Swedish Environmental Protection Agency and the US National Association of State Fire Marshals and corresponding USA environmental agencies jointly run the system. The goal of Green Flame is to promote the design of products and systems that are fire resistant but environmentally safe.

The Swedish National Association of State Fire Marshals, a member of Green Flame, has been at the forefront of advocating more functionbased alternatives to BFRs They apply their expertise in fire prevention in buildings to that of product redesign. They emphasise the role of product design in avoiding flammability and advocate that such criteria should be part of eco-labels. Prevention is key, and fires need to be detected at an early stage. They advocate the use of fault detectors to shut off the electrical supply, as well as automatic extinguishing systems inside personal computers and televisions. In particular, they point out sprinkler systems are used for protection inside racing cars so their use in personal cars, buses and trains is also possible and would provide a significant increase in fire safety. They also outline a selection of materials that can favour fire safety in certain applications:

- wood (to replace borders and edgings made of polymers)
- metal

- Glass, stone, and ceramics
- Plaster (to replace fake stucco details)
- leather, wool, cotton, linen, hemp
- living trees or flowers (to replace synthetic variants)
- stone or glass wool as insulation
- paper (packing materials). Flammable materials, air and a high temperature are needed to keep a fire burning. The State Fire Marshals advocate.
- lowering the energy that might be produced in the event of electrical failure
- lowering the power requirements for electrical appliances
- preventing heat production from friction in engines or movable parts in fans
- placing electrical heat sources at sufficient distances from flammable materials
- using internal fuses to cope with overloads or short circuits
- maintaining sufficient distances between warm parts and flammable materials
- cooling down parts which generate heat
- using an extinguishing system that reduces the temperature. Supply of air can be stopped by:
  - enclosing the heat producing technical components of products in metal.
  - using materials that creates a layer of tar (or similar) on burning.
  - designing sandwich constructions with non-inflammable surfaces.
  - using an extinguishing system that separates the fuel from the air.

The formation of a two year research programme into fire prevention within the EU will accelerate the move to safer substitutes to BFRs in buildings. The project, backed by the Commission, began in 2003. The aim is to stage a series of workshops at which member States can share expertise and experience as a basis for agreeing best pract ce in fire prevention. Substituting lead in electronic products The WEEE Directive was a major catalyst for research and adoption of lead free solutions in electronic equipment. The range of alternatives to lead in soldering range from tin (Sn), silver (Ag), copper (Cu), bismuth (Bi) and zinc (Zn). These heavy metals do not have the persistence and bioaccumulation pote pagers by December 1998; the use of Sn-Zn- Bi in their notebooks PCs by October 1999; the use of Sn-Ag-Cu in their main computers and equipment by 2002. Fujitsu targeted the use of Sn-Ag-Cu for their high end servers by October 1999; the use of Sn-Bi-Ag for their main board; and all new products to use Sn-Ag-Cu, Sn-Bi-Ag by the end of 2002.

Sony targeted the use of Sn-Ag-Bi-Cu for use in their digital video cameras by March 2000; and all products, electronic components and maintenance services to be lead free by end of March 2006 Hitachi targeted the use of Sn-2Ag-2Bi-0.5Cu in video cameras, vacuum cleaners, and washing machines by spring 1999; and all new products will use Sn-Ag-Cu, Sn0Bi+Ag- Cu(+Bi/In) by March 2002.

Matsushita targeted the use of Sn-Ag0Bi in mini disks by September 1998; and all new products to use Sn-Ag-Cu, Sn-Ag-Bi by March 2003. Panasonic achieved the full adoption of Pbfree solder using Sn-Cu in 2001. Philips has developed new lighting for cars – the Philips HiPerVision Technology which provides lighting for the automotive industry and uses 99% less lead.

Retailers are substituting a range of hazardous materials:

- Phthalate esters are used as softeners in flexible PVC products, including floors, wallpapers, furnishings, clothing and toys, as well as ingredients in cosmetics and perfumes.
- Organotin compounds are used as stabilisers in plastics, especially PVC, and tributyltin (TBT) is used as a treatment against dust mites and mould in some carpets and PVC floorings.
- Alkylphenols and their derivatives are primarily used as non ionic surfactants in industrial detergents, though also used in textile and leather finishing treatments, water based paints and as components of some personal care products.
- Artificial musks are used in fragrances and perfumes.

Boots decided to phase out alkylphenols in 1999 and by 2000 they had achieved 90% phase out. A difficulty was first identifying where alkylphenol ethoxylates (APEs) were used in their 40,000-50,000 product range as some of their supply chains were complex. Also, finding substitutes for some products proved problematic on quality grounds, for example, all artificial musks in cosmetic lines were phased out but some substitutes did not hold the fragrance for as long. In general, they have substituted alcohol ethoxylates as replacements for APEs. According to Boots, unlike APEs, alcohol ethoxylates appear to be readily and completely broken down in the environment Marks and Spencer initially targeted the successful phase out of pesticides in their product lines and then turned their attention to PVC. By eliminating PVC, they effectively eliminated most phthalates. Their transition from PVC in handbags, belts and shoes led to significant cost increases, but they have achieved a full phase out as well as an 89% phase out of PVC in packaging. Alternative plastics such as PET were used in some instances. Marks and Spencer targeted the substitution of alkyl tins in the dyeing and finishing of clothing, along with azo dyes and APEs some years ago. Their product specialists are working with the Green Chemistry department at York University to explore safer alternatives. They have not yet found substitutes for bisphenol A in some of their tin can linings. They observed that the market needs to move en masse with this issue and that major multinationals such as Coca Cola, Heinz and Walmart could demand safer substitutes if they chose to do so. Safeway phased out musk xylenes and nitro musks by reformulating their product line.

They also achieved a phase out of phthalates in fragrances in all but one line, and are working to eliminate this as well. They seek information from the International Fragrance Research Association, which has developed guidelines on what is permitted.

Homecare Products got their suppliers to remove a polycyclic musk and diethyl phthalates from their products after talking to Greenpeace in August 2001. Their suppliers subsequently informed them that the two substances added no benefit to the cleaning properties of the product. Other than a phase out of all BFRs, H&M have achieved a strict phase out of APEs, organotins, azo dyes and all carcinogenic dyes, PVC, bisphenol A, phthalates, antimony and a wide range of heavy metals nd chlorinated aromatic hydrocarbons.

They stipulated a clear set of criteria to all their suppliers, used testing to ensure compliance and relied on their suppliers and chemical formulators to provide alternatives.

The Body Shop has a set of guidelines that are used when formulating new products. If a chemical is of concern (and this concern has some justification), then it is either prohibited from use in new formulations or it is restricted to certain product types (leave-on versus rinse-off, for example). he Body Shop does not use PVC in its packaging or its accessory products and in this way avoids the use of higher phthalates – diethylhexyl phthalate and diisononyl phthalate. There is a range of alternative plastics available. Many cosmetics manufacturers have used the lower phthalates (diethyl phthalate, dimethyl nd dibutyl phthalates) for many years as solvents and diluents in perfumes. The Body Shop has also used phthalates for this application. The Body Shop states that while there is no clear evidence that they are responsible for any hormone disruption, as a precautionary measure they have decided to avoid the use of all phthalates in new perfumes, and aim to phase out phthalates that remain in existing perfumes. There are alternative solvents available such as dipropylene glycol, carbitol and ethyldiglycol. On the whole, the cosmetic industry is able to innovate quickly because cosmetic products have a relatively short life cycle; this provides short to medium term opportunities to formulate out the problem ingredients in the next version of the product. Raw materials come at different costs and reformulating out chemicals of concern may result in an increase in material cost. However, as more companies switch to alternatives, economies of scale are realised and the price will fall. Colgate, Palmolive and Procter and Gamble have removed APEs from all or part of their personal care ranges Nike, Lego, Mattel, and Sony are some of the companies that have phased out PVC plastic. Ford, Peugeot, Daimler Benz, Opel, Volkswagen, BMW, Mercedes Benz, Mitsubishi, Nissan, and Toyota are all adopting PVC restrictions PVC plastic is the largest end use of phthalates and a significant end use for organotins and BFRs with antimony. A phase out of PVC therefore directly reduces the use of these hazardous constituents. Alternatives to PVC vary and must be assessed on their chemical profile. Greenpeace has charted the move to PVC alternatives over the last few years and has compiled an extensive database of PVC Restrictions World Wide which is available on line. Chemical suppliers are adopting Green Chemistry The chemical industry is often the strongest link in the supply chain since they provide the chemical formulations to their buyers and are in control of research into alternatives. The recent emergence of Green Chemistry and its 12 criteria has spurred innovation into safer chemicals. A few examples illustrate the progress that is taking place and suggest a wider potential for innovation if a greater market demand for safer chemicals was to occur.

Pfizer has implemented a substitution for a range of hazardous solvents (methylene chloride, tetrahydrofuran, toluene, hexane) in its formulation of Zoloft – the most prescribed agent of its kind to treat depression.

The new synthesis involves optimising the more benign solvent, ethanol, in its process and has achieved significant hazard waste reductions as part of its conversion. Bayer has developed a new chelating agent, which is benign from a toxicological and ecotoxicological standpoint, in a 'waste free and environmentally friendly manufacturing process'. The chemical, sodium iminodisuccinate, belongs to the aminocarboxylate class of chelating agents. Nearly all this chemical in use today is produced from amines,formaldehyde, sodium hydroxide and hydrogen cyanide. The industrial use of thousands of tons of hydrogen cyanide is an extreme toxicity hazard. In contrast, Bayer's substitute is produced from maleic anhydride, sodium hydroxide and water. The only solvent used in the production process is water and the only by-product formed – ammonia dissolved in water – is recycled back into the product cycle. PPG Industries has developed a substitute for lead in a coating process used widely in the auto industry. The replacement is yttrium which, although much less studied than lead, is considered orders of magnitude lower in hazard. In addition it was discovered that as yttrium is used in the process it is converted to yttrium oxide which is appears to be nontoxic by ingestion, in stark contrast to lead.

As PPG customers implement yttrium over the next several years, it is projected that the use of approximately one million pounds of lead will be avoided.

Solutia (formerly Monsanto) one of the world's largest producers of chlorinated aromatics, has funded research over the years to explore alternative synthetic reactions for manufacturing processes that do not require the use of chlorine. They have found a nonhalogen route for the manufacture of a key intermediate in the rubber chemicals family of antidegradants. The new process generates 74% less organic waste, 99% less inorganic waste and 97% less wastewater. The use of carbon dioxide (CO_2) as a replacement for halogenated solvents has long been recognised as an ideal solvent that is nontoxic, non-flammable, safe to work with and reusable. A new CO_2 surfactant system, discovered by a Professor DeSimone at the University of North Carolina, will expand the use of CO_2 as an alternative to chlorinated solvents currently used in manufacturing and garment care industries. The use of CO_2 as a blowing agent in polystyrene has now been developed by Dow Chemical Company as a replacement for ozone depleting CFC-12.

Political initiatives on substitution In 1998 at a meeting held in Sintra, OSPAR delegates from each of the 15 States of the North East Atlantic Region and the inate releases of hazardous substances into the marine environment by the year 2020. As a first step towards implementing this goal, OSPAR agreed on a 'List of Chemicals for Priority Action', a list of 15 chemicals which would be dealt with by 'the drawing up of programs and measures by 2003 for the control of discharges, emissions and losses of substances on [the Priority] list, and their substitution with less hazardous or non-hazardous substances where feasible.'

The OSPAR Commission invited industry to help achieve this objective through the incorporation of clean production and clean products and the development of less hazardous or preferably non-hazardous substances. This focus on substitution and the reference to industry developments has been mirrored elsewhere. This annex provides an overview of existing EU law, mostly directives, which incorporate the principle of substitution. In the text, substitution is generally referred to as replacing a substance of concern with a safer alternative. All the highlighted directives and the European Court Ruling, as well as the two examples of internationally binding agreements that mandate substitution, have been analysed closely with respect to the scope of each text and the rationale for the requirement of substitution. For this purpose it seemed most practical to present the findings in a table format.

9

Atom Economy

A main focus of green chemistry is to reduce the amount of pollution created in chemical processes. Reactions in which a large proportion of the reactant atoms end up in waste products contribute to pollution, make ineffective use of resources, and raise the costs of production. While percent yield is often considered a measure of how efficiently reactants are used in making a final product, it neglects to measure what fraction of the reactant atoms end up in the desired product. Atom economy, a concept introduced by Barry Trost of Stanford University, is a ratio that lets us know what percent of the reactants' atoms make their way into the desired product.

Consider the Following Statement

A chemist suggests that it is impossible to "throw away" anything. The chemist goes on to say there is no such thing as "waste." Obviously, we dispose of many items in our everyday lives. What do you suppose this chemist means by saying says things can't be thrown away?

What is a "waste" product?

1. The Law of Conservation of Matter states "matter cannot be created or destroyed." How does this law relate to the statement made above by the chemist? Atoms don't go away just because they have reacted, even though uninformed observations sometimes suggest this. When a candle burns, matter appears to be reduced and energy is given off. In fact, the matter simply turned into less observable gases and smoke.
2. Chemists sometimes refer to "byproducts" rather than "waste." What are some advantages and disadvantages of using such a term? List one of each. A byproduct is still a product, so it helps reinforce the idea that "other stuff" was produced and eeds to be considered. A disadvantage is that the term byproduct may suggest that it is useful, when it may not be.
3. On aver age, you have a couple of atoms in your body that were once found in the body of a dinosaur. Explain how this can be Atoms are continually recycled. A carbon atom that is part of your protein today will eventually be returned to the environment, where it will be incorporated in new compounds, perhaps in another living organism. Given the quantities of carbon atoms required to make a dinosaur and a human, it is statistically likely that some of your atoms were once used to make a dinosaur.

Instructional Notes

This activity is designed to reinforce the students' understanding of the Law of Conservation of Matter. All substances that we consider waste are made up of molecules and atoms that do not disappear just because we dispose of them. Things we label as waste and throw away because we have no use for them can be put out of sight and out of mind—but not out of existence. In a chemical reaction, atoms may end up as part of either the desired product or undesired byproducts.

It is good for students to realize that the number and kinds of atoms on earth are essentially fixed. A small amount of matter might leave the earth, a bit is gained from meteorites and such, and we have a limited amount of transmutation due to nuclear reactions. But other than that, what we have is what we have.

Student misconceptions might arise from the observation of matter changing so profoundly in chemical reactions. Still, the fact is that chemical reactions only involve the making and breaking of bonds between atoms. The kind and total number of atoms remain essentially the same.

Counting Atoms

How can we calculate the atom economy of a simple chemical reaction? (Activity) Chemists can build molecules by knowing the rules about how atoms combine. Chemical bonds being broken and new bonds forming can explain all chemical reactions. One simple organic molecule is propene, which is composed of three carbon atoms and six hydrogen atoms and has one double bond between two of the carbon atoms. There is more than one way to make propene. In this activity, you will consider two different reactions leading to propene and will use molecular model kits to investigate how the concept of atom economy can illustrate green chemistry.

1. Use your molecular model kit (TinkerToys) to construct a trimethlypropyl amine ion. When this substance is heated, it decomposes into propene and two other products as shown below. Count the number of each kind of atom in the molecule.

2. "Heat" will be applied to your model and you will break it apart. Make one propene molecule. Place the rest of the atoms in a box marked "waste products." Count the number and kind of atoms in the propene molecule and the number and kind of atoms in the "waste box". When you are finished, dismantle your molecules.

3. Use your model kit to make a 1-propanol molecule. When this molecule is heated with a sulfuric acid catalyst, it also forms propene.

4. When your teacher indicates, break your molecule apart and make one propene molecule. Place the rest of the atoms in a box marked "waste products." Count the number and kind of atoms in the propene molecule and the number and kind of atoms in the "waste box." When you are finished, dismantle your molecules.

Counting Atoms

1. Compare the formula weight of the atoms in the waste box in reaction:

   – to the formula weight of the atoms in the waste box in reaction;

   – While this is not the way propene is produced on a mass scale, suppose reaction 1 were the conventional way of producing propene. What is the atom economy of this reaction?

Note: To quantify the concept of atom economy, we can calculate "percent atom economy". To do this, we divide the formula weight of the desired product by the formula weight of all the products of the reaction and multiply the result by 100. The calculation is displayed below: Formula weight of desired product X 100 = Percent atom economy Formula weight of all products

Reaction 1 (Fig. 9.1): 42g desired product, 119g total of all products; atom economy = 35%

Reaction 2 (Fig. 9.2): 42g desired product, 60g total of all products; atom economy = 70%. What advantages would there be if reaction 2 were offered as a green chemistry alternative for the production of propene? What is the percent atom economy of this reaction? Not only is the atom economy double in reaction 2, but the other product is simply water, a very benign byproduct.

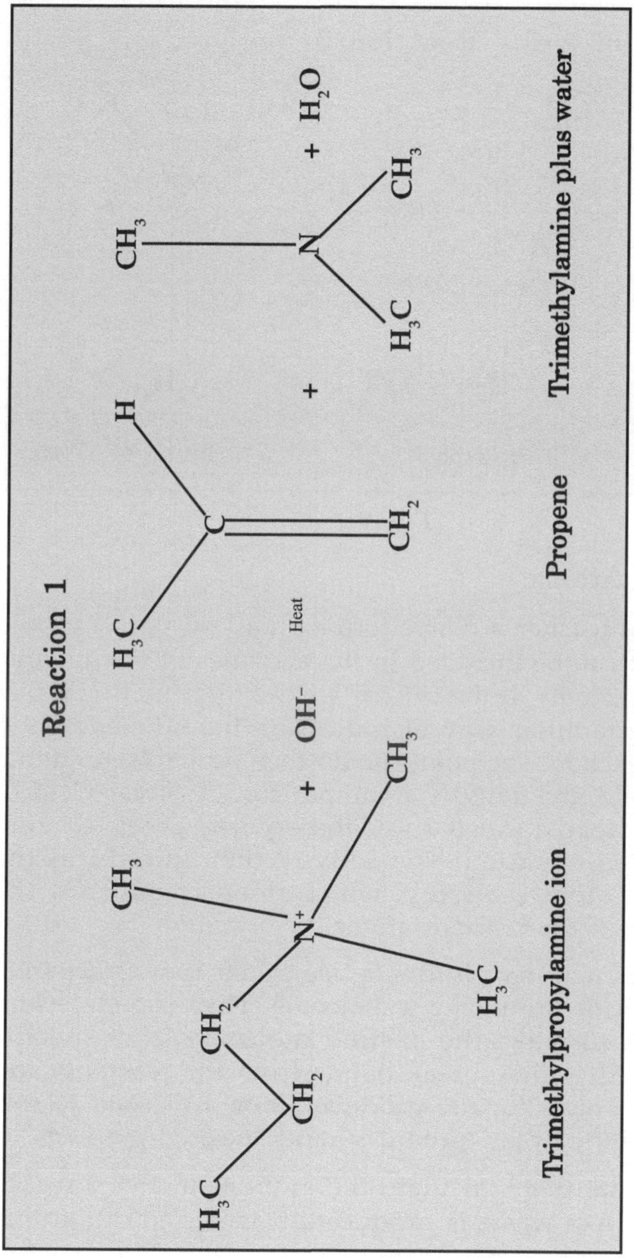
Reaction 1
Heat
Trimethylpropylamine ion
Propene
Trimethylamine plus water

Fig. 9.1

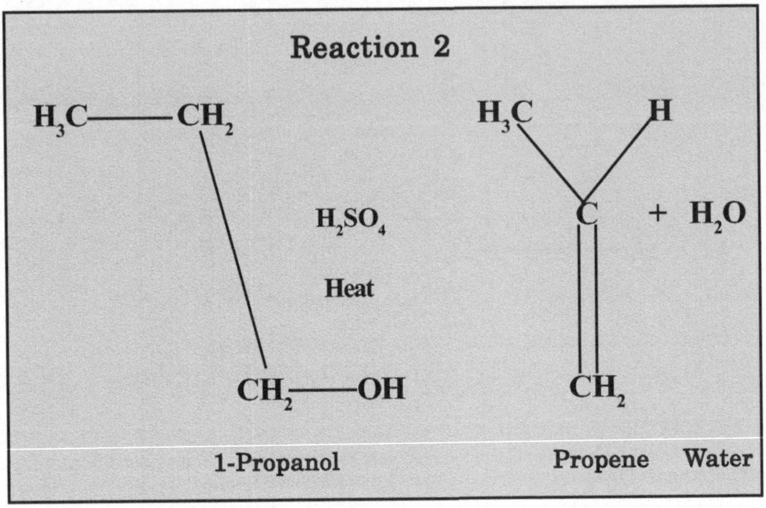

Fig. 9.2

Counting Atoms

At the teacher's discretion, a standard percent yield experiment can be included in the activities of this module before this activity, if the teacher wishes to demonstrate that this is one familiar way of indicating the efficiency of a chemical reaction. The point should then be emphasized that percent yield generally is a comparison of actual yield of product compared to what was theoretically predicted from the balanced equation. The activity then introduces the concept of atom economy, which does account for the destination of the reactant atoms.

In this activity, students are given two alternative procedures for producing a molecule. They use molecular model kits to make the desired molecule from specific reactants. After they break bonds from the reactants and make new bonds for the products, they are asked to sort the atoms into "waste products" and "desired products".

Students then calculate the atom economy for each reaction. If less waste is produced, it is an indication that the process might be more "green." Of course, a more acceptable atom economy is not sufficient on its own to make a process a good example of green chemistry. If the waste were more toxic than that which is produced in conventional processes, the reaction would be unacceptable. Likewise, the new process has to make sense economically. Use whatever kind of molecular modeling kits you have available. You may have commercial model kits, TinkerToys, or alternatively, you can use gumdrops or other candies plus toothpicks to make the models. Any of these options will work well. Atom economy in the real world—Ibuprofen. (Fig. 9.3) (Discussion), In the Counting Atoms activity, you modeled chemical reactions and placed "unused" atoms into a "waste box." In chemical reactions, atoms that are not in the desired product are considered waste. This waste is analogous to the waste box in the previous activity. To make complex molecules (like the ones used in pharmaceuticals and pesticides), chemical engineers will often link together several reactions to form a chemical synthesis. Each step in the synthesis can be viewed as an individual reaction. As you saw in the Counting Atoms activity, atoms in a reaction can either end up in the final product or be discarded in a waste box.

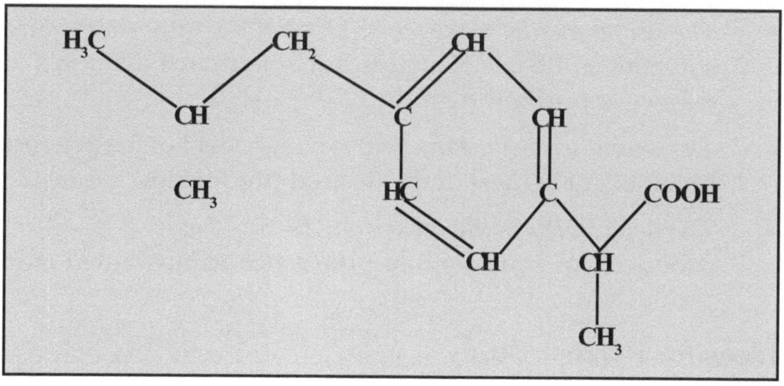

Fig. 9.3
Ibuprofen

Let's look at a common synthesis of ibuprofen (the active ingredient in the over-the-counter drugs Advil and Motrin, see below for its chemical structure). See the pages following this activity for the synthesis (Boots Co. synthesis from the 1960s)—you may display to your students on an overhead. The unused atoms during the synthesis are analogous to the atoms that you placed in the waste box in the Counting Atoms activity. In the early 1990s, the BHC Co. developed a new synthesis for ibuprofen that created far less waste because fewer atoms made their way into the waste box. Discussing this example with your students provides a "real world" connection for the concept of atom economy.

Atom economy is a useful concept for determining the efficiency of use of starting materials in reactions orsyntheses. However, the presence of atom economy is not sufficient to consider a process "green." Some other important aspects of the process that must also be considered include:

- *The nature of the waste produced:* Is the waste toxic or environmentally harmful?
- *The amount of energy needed to ake the reaction proceed:* Does the reaction require excessive amounts of energy?
- *The use of auxiliary reagents:* Does the synthesis require solvents or the use of significant amounts of materials to extract or purify the product?
- *The percent yield of the reaction:* Is a high yield of the product obtained, or is most of the desired product lost as waste?
- *The selectivity of the reaction:* Is the desired product produced in considerable preference to unwanted side products?

Ideas for Further Study

As an extension of your discussion about what a "waste" product is, students may benefit from considering what happens to the "waste" that we throw out each day.

Where does "garbage" go in your community? Is there a local landfill, or is the waste transported to landfills elsewhere? What happens to materials once it is placed in a landfill? Do local recycling and composting programs exist? See the supplemental activities on solid waste and packaging that follow this activity for additional exercises that may be used as extensions of this exercise. Another useful extension of the discussion of wastes is to perform a Life Cycle Analysis (LCA) of an everyday item. See the following Waste Management and Research Center (WMRC) fact sheet for an activity involving the LCA of a tennis shoe. Finally, a great way to convey the idea of process efficiency to younger students is use a game as an analogy. One game that illustrates the imprtance of efficient processes graphically is the "peg game that is commonly found at Cracker Barrel restaurants. The object of this game is to move your pieces in such a way as to only have one peg remaining. If you don't move efficiently, you'll have additional pegs standing at the end of the game (which can be thought of as "wastes") and you lose.

A quarter of the world's people have to survive on less than 70p a day. Millions have no health care and the world's population is expected to increase by about another 3 billion over the next 50 years. Even in developed nations, poverty, education and healthcare could be improved.

To help deal with this situation, the world's economy needs to grow; in particular, the economies of developing nations need to expand. However, economic growth is often linked to environmental pollution problems. The challenge is to develop in a way that meets the needs of the present generation without compromising the ability of future generations to meet their own needs – in other words, without causing a lot of environmental damage and wasting limited resources. This type of development is called 'sustainable development' and it will be more and more critical as the population of the world increases.

One of the ways in which the chemical industry is working towards sustainable development is by using 'Green Chemistry.' One of the basic ideas of Green Chemistry is to prevent pollution and the production of hazardous materials instead of producing them and then cleaning them up.

This means that Green Chemistry: Is safe Conserves raw materials and energy Is more cost effective than conventional methods.

Yield

Most of the chemical industry is concerned with turning one material (the raw material) into another one that is more useful and valuable (the product). This process may have several steps and is called a 'chemical synthesis.' All the designers of chemical processes want to make the maximum amount of product they can from a given raw material. It is possible to calculate how successful one of these processes is by using the idea of yield.

% Yield =

$$\frac{\text{Mass of product actually made}}{\text{Maximum mass product that could be made}} \times 100$$

1. In a chemical synthesis of calcium oxide, calcium carbonate is roasted in an oven. The equation for the reaction is:

   $CaCo_3 = CaO + CO_2$

   RMM or Mr: 100 56 44 which shows the maximum mass that could be made If 50 kg calcium carbonate is used and 21 kg calcium oxide is made, what is the percentage yield of the reaction?

2. To make sulfur dioxide, sulfur is burnt in oxygen:

   $S + O_2 = SO_2$

If 8 g sulfur is used, what is the maximum mass of sulfur dioxide that could be produced? If 14 g sulfur dioxide is produced, what is the percentage yield?

Atom Economy

The idea of yield is useful, but from a Green Chemistry and sustainable development perspective, it is not the full picture. This is because yield is calculated by considering only one reactant and one product. One of the key principles of Green Chemistry is that processes should be designed so that the maximum amount of all the raw materials ends up in the product and a minimum amount of waste is produced. A reaction can have a high percentage yield but also make a lot of waste product. This kind of reaction has a low atom economy. Both the yield and the atom economy have to be taken into account when designing a green chemical process. Look again at the reaction you considered in Q_3:

$CaCO_3 = CaO + CO_2$

RMM or Mr: 100 56 44

If we split up the formulae, we can look at what happens to each atom in the reaction. The atoms shown below in bold end up in the product we want, the rest do not:

Ca C O O O = Ca O + C O O

10

Catalysis

Catalysis is the process in which the rate of a chemical reaction is either increased or decreased by means of a chemical substance known as a catalyst. Unlike other reagents that participate in the chemical reaction, a catalyst is not consumed by the reaction itself. The catalyst may participate in multiple chemical transformations. Catalysts that speed the reaction are called positive catalysts. Catalysts that slow down the reaction are called negative catalysts or inhibitors. Substances that increase the activity of catalysts are called promoters and substances that deactivate catalysts are called catalytic poisons. For instance, in the reduction of ethyne to ethene, the catalyst is palladium (Pd) partly "poisoned" with lead(II) acetate [$Pb(CH_3COO)_2$]. Without the deactivation of the catalyst, the produced ethene will be further reduced to ethane.

The general feature of catalysis is that the catalytic reaction has a lower rate-limiting free energy change to the transition state than the corresponding uncatalyzed reaction, resulting in a larger reaction rate at lower temperature.

However, the mechanistic origin of catalysis is complex. Catalysts may affect the reaction environment favorably, e.g. acid catalysts for reactions of carbonyl compounds, form specific intermediates that are not produced naturally, such as osmate esters in osmium tetroxide-catalyzed dihydroxylation of alkenes, or cause lysis of reagents to reactive forms, such as atomic hydrogen in catalytic hydrogenation.

Kinetically, catalytic reactions behave like typical chemical reactions, i.e. the reaction rate depends on the frequency of contact of the reactants in the rate-determining step. Usually, the catalyst participates in this slow step, and rates are limited by amount of catalyst. In heterogeneous catalysis, the diffusion of reagents to the surface and diffusion of products from the surface can be rate determining. Analogous events associated with substrate binding and product dissociation apply to homogeneous catalysts.

Although catalysts are not consumed by the reaction itself, they may be inhibited, deactivated or destroyed by secondary processes. In heterogeneous catalysis, typical secondary processes include coking where the catalyst becomes covered by polymeric side products. Similarly heterogeneous catalysts can dissolve or evaporate into the mobile medium, solution or gas-phase. These side reactions may even include following elementary reaction steps, as in alkali-catalyzed hydrolysis of esters, thus requiring a stoichiometric amount of catalyst.

Background

The production of most industrially important chemicals involves catalysis. Similarly, most biochemically significant processes are catalysed. Research into catalysis is a major field in applied science and involves many areas of chemistry, notably in organometallic chemistry and materials science. Catalysis is relevant to many aspects of environmental science, e.g. the catalytic converter in automobiles and the dynamics of the ozone hole. Catalytic reactions are preferred in environmentally friendly green chemistry due to the reduced amount of waste generated as opposed to stoichiometric reactions in which all reactants are consumed and more side products are formed. The most common catalyst is the proton (H^+). Many transition metals and transition metal complexes are used in catalysis as well. Catalysts called enzymes are important in biology.

A catalyst works by providing an alternative reaction pathway to the reaction product. The rate of the reaction is increased as this alternative route has a lower activation energy than the reaction route not mediated by the catalyst. The disproportionation of hydrogen peroxide to give water and oxygen is a reaction that is strongly affected by catalysts:

$$2\ H_2O_2 = 2\ H_2O + O_2$$

This reaction is favoured in the sense that reaction products are more stable than the starting material, however the uncatalysed reaction is slow. The decomposition of hydrogen peroxide is in fact so slow that hydrogen peroxide solutions are commercially available. Upon the addition of a small amount of manganese dioxide, the hydrogen peroxide rapidly reacts according to the above equation. This effect is readily seen by the effervescence of oxygen The manganese dioxide may be recovered unchanged, and re-used indefinitely, and thus is not consumed in the reaction. Accordingly, manganese dioxide *catalyses* this reaction.

General Principles of Catalysis

Typical Mechanism

Catalysts generally react with one or more reactants to form intermediates that subsequently give the final reaction product, in the process regenerating the catalyst. The following is a typical reaction scheme, where *C* represents the catalyst, X and Y are reactants, and Z is the product of the reaction of X and Y:

$$X + C = XC \tag{10.1}$$

$$Y + XC = XYC \tag{10.2}$$

$$XYC = CZ \tag{10.3}$$

$$CZ = C + Z \tag{10.4}$$

Although the catalyst is consumed by reaction 1, it is subsequently produced by reaction 4, so for the overall reaction:

$$X + Y = Z$$

As a catalyst is regenerated in a reaction, often only small amounts are needed to increase the rate of the reaction. In practice, however, catalysts are sometimes consumed in secondary processes.

As an example of this process, in 2008 Danish researchers first revealed the sequence of events when oxygen and hydrogen combine on the surface of titanium dioxide (TiO_2, or *titania*) to produce water. With a time-lapse series of scanning tunneling microscopy images, they determined the molecules undergo adsorption, dissociation and diffusion before reacting. The intermediate reaction states were: HO_2, H_2O_2, then H_3O_2 and the final reaction product (water molecule dimers), after which the water molecule desorbs from the catalyst surface.

Catalysis and Reaction Energetics

Catalysts work by providing an (alternative) mechanism involving a different transition state and lower activation energy. Consequently, more molecular collisions have the energy needed to reach the transition state. Hence, catalysts can enable reactions that would otherwise be blocked or slowed by a kinetic barrier. The catalyst may increase reaction rate or selectivity, or enable the reaction at lower temperatures. This effect can be illustrated with a Boltzmann distribution and energy profile diagram.

Catalysts do not change the extent of a reaction: they have no effect on the chemical equilibrium of a reaction because the rate of both the forward and the reverse reaction are both affected. The fact that a catalyst does not change the equilibrium is a consequence of the second law of thermodynamics. Suppose there was such a catalyst that shifted an equilibrium. Introducing the catalyst to the system would result in reaction to move to the new equilibrium, producing energy. Production of energy is a necessary result since reactions are spontaneous if and only if Gibbs free energy is produced, and if there is no energy barrier, there is no need for catalyst. Then, removing the catalyst would also result in reaction, producing energy; i.e. the addition and its reverse process, removal, would both produce energy. Thus, a catalyst that could change the equilibrium would be a perpetual motion machine, a contradiction to the laws of thermodynamics.

If a catalyst does change the equilibrium, then it must be consumed as the reaction proceeds, and thus it is also a reactant. Illustrative is the base-catalysed hydrolysis of esters.

The SI derived unit for measuring the catalytic activity of a catalyst is the katal, which is moles per second. The activity of a catalyst can also be described by the turn over number (or TON) and the catalytic efficiency by the *turn over frequency* (TOF). The biochemical equivalent is the enzyme unit. For more information on the efficiency of enzymatic catalysis, see the article on Enzymes.

The catalyst stabilizes the transition state more than it stabilizes the starting material. It decreases the kinetic barrier by decreasing the *difference* in energy between starting material and transition state.

Typical Catalytic Materials

The chemical nature of catalysts is as diverse as catalysis itself, although some generalizations can be made. Proton acids are probably the most widely used catalysts, especially for the many reactions involving water, including hydrolysis and its reverse. Multifunctional solids often are catalytically active, e.g. zeolites, alumina and certain forms of graphitic carbon. Transition metals are often used to catalyse redox reactions (oxidation, hydrogenation). Many catalytic processes, especially those involving hydrogen, require platinum metals.

Some so-called catalysts are really "precatalysts." Precatalysts convert to catalysts in the reaction. For example, Wilkinson's catalyst $RhCl(PPh_3)_3$ loses one triphenylphosphine ligand before entering the true catalytic cycle. Precatalysts are easier to store but are easily activated in situ. Because of this preactivation step, many catalytic reactions involve an induction period.

Types of Catalysis

Catalysts can be either heterogeneous or homogeneous, depending on whether a catalyst exists in the same phase as the substrate. Biocatalysts are often seen as a separate group.

Heterogeneous Catalysts

Heterogeneous catalysts are those which act in a different phases than the reactants. Most heterogeneous catalysts are solids that act on substrates in a liquid or gaseous reaction mixture. Diverse mechanisms for reactions on surfaces are known, depending on how the adsorption takes place (Langmuir-Hinshelwood and Eley-Rideal)

For example, in the Haber process, finely divided iron serves as a catalyst for the synthesis of ammonia from nitrogen and hydrogen. The reacting gases adsorb onto "active sites" on the iron particles. Once adsorbed, the bonds within the reacting molecules are weakened, and new bonds between the resulting fragments form in part due to their close proximity. In this way the particularly strong triple bond in nitrogen is weakened and the hydrogen and nitrogen atoms combine faster than would be the case in the gas phase, so the rate of reaction increases.

Heterogeneous catalysts are typically "supported", which means that the catalyst is dispersed on a second material that enhances the effectiveness or minimizes their cost. Sometimes the support is merely a surface upon which the catalyst is spread to increase the surface area. More often, the support and the catalyst interact, affecting the catalytic reaction.

Homogeneous Catalysts

Homogeneous catalysts function in the same phase as the reactants, but the mechanistic principles invoked in heterogeneous catalysis are generally applicable. Typically homogeneous catalysts are dissolved in a solvent with the substrates. One example of homogeneous catalysis involves the influence of H^+ on the esterification of esters, e.g. methyl acetate from acetic acid and methanol.For inorganic chemists, homogeneous catalysis is often synonymous with organometallic catalysts.

Electrocatalysts

In the context of electrochemistry, specifically in fuel cell engineering, various metal-containing catalysts are used to enhance the rates of the half reactions that comprise the fuel cell. One common type of fuel cell electrocatalyst is based upon nanoparticles of platinum that are supported on slightly larger carbon particles. When this platinum electrocatalyst is in contact with one of the electrodes in a fuel cell, it increases the rate of oxygen reduction to water (or hydroxide or hydrogen peroxide).

Organocatalysis

Whereas transition metals sometimes attract most of the attention in the study of catalysis, organic molecules without metals can also possess catalytic properties. Typically, organic catalysts require a higher loading (or amount of catalyst per unit amount of reactant) than transition metal-based catalysts, but these catalysts are usually commercially available in bulk, helping to reduce costs. In the early 2000s, organocatalysts were considered "new generation" and are competitive to traditional metal-containing catalysts. Enzymatic reactions operate via the principles of organic catalysis.

Significance of Catalysis

Estimates are that 90% of all commercially produced chemical products involve catalysts at some stage in the process of their manufacture. In 2005, catalytic processes generated about $900 billion in products worldwide. Catalysis is so pervasive that subareas are not readily classified.

Energy Processing

Petroleum refining makes intensive use of catalysis for alkylation, catalytic cracking (breaking long-chain hydrocarbons into smaller pieces), naphtha reforming and steam reforming (conversion of hydrocarbons into synthesis gas). Even the exhaust from the burning of fossil fuels is treated via catalysis: Catalytic converters, typically composed of platinum and rhodium, break down some of the more harmful byproducts of automobile exhaust.

$$2\ CO + 2\ NO = 2\ CO_2 + N_2$$

With regards to synthetic fuels, an old but still important process is the Fischer-Tropsch synthesis of hydrocarbons from synthesis gas, which itself is processed via water-gas shift reactions, catalysed by iron. Biodiesel and related biofuels require processing via both inorganic and biocatalysts. Fuel cells rely on catalysts for both the anodic and cathodic reactions.

Bulk Chemicals

Some of the largest-scale chemicals are produced via catalytic oxidation, often using oxygen. Examples include nitric acid (from ammonia), sulfuric acid (from sulfur dioxide to sulfur trioxide by the chamber process), terephthalic acid from p-xylene, and acrylonitrile from propane and ammonia.

Many other chemical products are generated by large-scale reduction, often via hydrogenation. The largest-scale example is ammonia, which is prepared via the Haber process from nitrogen. Methanol is prepared from carbon monoxide.

Bulk polymers derived from ethylene and propylene are often prepared via Ziegler-Natta catalysis. Polyesters, polyamides, and isocyanates are derived via acid-base catalysis.

Most carbonylation processes require metal catalysts, examples include the Monsanto acetic acid process and hydroformylation.

Fine Chemicals

Many fine chemicals are prepared via catalysis; methods include those of heavy industry as well as more specialized processes that would be prohibitively expensive on a large scale. Examples include olefin metathesis using Grubbs' catalyst, the Heck reaction, and Friedel-Crafts reactions. Because most bioactive compounds are chiral, many pharmaceuticals are produced by enantioselective catalysis.

Food Processing

One of the most obvious applications of catalysis is the hydrogenation (reaction with hydrogen gas) of fats using nickel catalyst to produce margarine. Many other foodstuffs are prepared via biocatalysis.

Biology

In nature, enzymes are catalysts in metabolism and catabolism. Most biocatalysts are protein-based, i.e. enzymes, but other classes of biomolecules also exhibit catalytic properties including ribozymes, and synthetic deoxyribozymes. Biocatalysts can be thought of as intermediate between homogenous and heterogeneous catalysts, although strictly speaking soluble enzymes are homogeneous catalysts and membrane-bound enzymes are heterogeneous. Several factors affect the activity of enzymes (and other catalysts) including temperature, pH, concentration of enzyme, substrate, and products.

A particularly important reagent in enzymatic reactions is water, which is the product of many bond-forming reactions and a reactant in many bond-breaking processes. Enzymes are employed to prepare many commodity chemicals including high-fructose corn syrup and acrylamide.

In the Environment

Catalysis impacts the environment by increasing the efficiency of industrial processes, but catalysis also plays a direct role in the environment. A notable example is the catalytic role of Chlorine free radicals in the breakdown of ozone. These radicals are formed by the action of ultraviolet radiation on chlorofluorocarbons (CFCs).

$$Cl + O_3 = ClO + O_2$$

$$ClO + O = Cl + O_2$$

History

In a general sense, anything that increases the rate of a process is a "catalyst", a term derived from Greek meaning "to annul," or "to untie," or "to pick up." The phrase *catalysed processes* was coined by Jöns Jakob Berzelius in 1836 to describe reactions that are accelerated by substances that remain unchanged after the reaction. Other early chemists involved in catalysis were Alexander Mitscherlich who referred to *contact processes* and Johann Wolfgang Döbereiner who spoke of *contact action* and whose lighter based on hydrogen and a platinum sponge became a huge commercial success in the 1820s. Humphry Davy discovered the use of platinum in catalysis. In the 1880s, Wilhelm Ostwald at Leipzig University started a systematic investigation into reactions that were catalyzed by the presence of acids and bases, and found that chemical reactions occur at finite rates and that these rates can be used to determine the strengths of acids and bases. For this work, Ostwald was awarded the 1909 Nobel Prize in Chemistry

Autocatalytic Reaction

Autocatalytic reactions are chemical reactions in which at least one of the products is also a reactant. The rate equations for autocatalytic reactions are fundamentally nonlinear. This nonlinearity can lead to the spontaneous generation of order. A dramatic example of this order is that which is found in living systems. This spontaneous order creation seems to contradict the Second Law of Thermodynamics. This contradiction is resolved when the disorder of both the system and its surroundings are taken into account.

Background

The Second Law of Thermodynamics states that the disorder (entropy) of a physical or chemical system and its surroundings must increase with time. In other words, systems left to themselves must become increasingly random. To say it yet another way, orderly energy of a system like uniform motion must degrade eventually to the random motion of particles in a heat bath.

This seems to run counter to experience. There are many instances in which physical systems spontaneously become emergent or orderly. For example, despite the destruction they do, hurricanes have a very orderly vortex motion when compared with the random motion of the air molecules in a closed room. Even more spectacular is the order created by chemical systems; the most dramatic being the order associated with life.

Our experience is consistent with the Second Law. The Second Law states that the total disorder of a system and its surroundings must increase with time. Order can be created in a system by an even greater decrease in order of the systems surroundings In the hurricane example, hurricanes are formed from unequal heating within the atmosphere. The Earth's atmosphere is then far from thermal equilibrium. The order of the Earth's atmosphere increases, but at the expense of the order of the sun. The sun is becoming more disorderly as it ages and throws off light and material to the rest of the universe. The total disorder of the sun and the earth increases despite the fact that orderly hurricanes are generated on earth.

A similar example exists for living chemical systems. The sun provides energy to green plants. The green plants are food for other living chemical systems. The energy absorbed by plants and converted into chemical energy generates a system on earth that is orderly and far from chemical equilibrium. Here, the difference from chemical equilibrium is determined by an excess of reactants over the equilibrium amount. Once again, order on earth is generated at the expense of entropy increase of the sun. The total entropy of the earth and the rest of the universe increases, consistent with the Second Law.

Not all chemical reactions, however, generate order. The class of reactions most closely associated with order creation is the class of autocatalytic reactions. These are reactions in which one or more of the products are the same as one or more of the reactants. Simple autocatalytic reactions (clock reactions) are known to oscillate in time, thus creating temporal order. Other simple reactions can generate spatial separation of chemical species generating spatial order. More complex reactions are involved in metabolic pathways and metabolic networks in biological systems.

The transition to order as the distance from equilibrium increases is not usually continuous. Order typically appears abruptly. The threshold between the disorder of chemical equilibrium and order is known as a phase transition. The conditions for a phase transition can be determined with the mathematical machinery of non-equilibrium thermodynamics.

Autocatalytic Reactions

Autocatalytic reactions are those in which at least one of the products is a reactant. Perhaps the simplest autocatalytic reaction can be written

$$A + B \rightleftharpoons 2B$$

with the rate equations

$$\frac{d}{dt}\{A\} = -k_+\{A\}\{B\} + k_-\{B\}^2$$

$$\frac{d}{dt}\{B\} = -k_+\{A\}\{B\} - k_-\{B\}^2$$

This reaction is one in which a molecule of species A interacts with a molecule of species B. The A molecule is converted into a B molecule. The final product consists of the original B molecule plus the B molecule created in the reaction.

The key feature of these rate equations is that they are nonlinear; the second term on the right varies as the square of the concentration of B. This feature can lead to multiple fixed points of the system, much like a quadratic equation can have multiple roots. Multiple fixed points allow for multiple states of the system. A system existing in multiple macroscopic states is more orderly (has lower entropy) than a system in a single state.

The Lotka-Volterra equation is isomorphic with the predator prey model and the two reaction autocatalytic model. In this example baboons and cheatahs are equivalent to two different chemical species in autocatalytic reactions.

CREATION OF ORDER

Temporal order

Idealized example: Lotka-Volterra equation

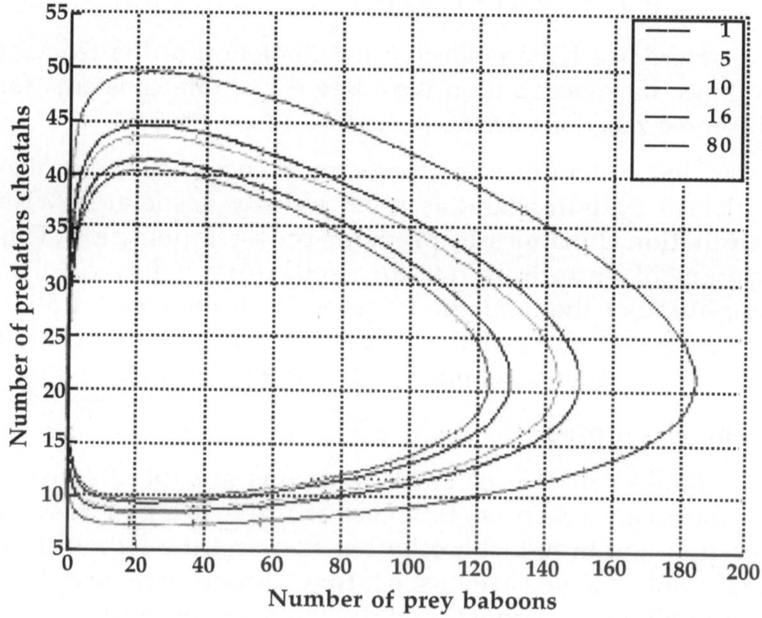

Fig. 10.1

Consider a coupled set of two autocatalytic reactions in which the concentration of one of the reactants A is much larger than its equilibrium value. In this case the forward reaction rate is so much larger than the reverse rates that we can neglect the reverse rates.

$$A + X \rightarrow 2X$$

$$X + Y \rightarrow 2Y$$

$$Y \rightarrow E$$

with the rate equations

$$\frac{d}{dt}\{X\}=k_1\{A\}\{X\}-k_2\{X\}\{Y\}$$

$$\frac{d}{dt}\{Y\}=k_2\{A\}\{X\}-k_3\{Y\}$$

Here, we have neglected the depletion of the reactant A, since its concentration is so large. The rate constants for the three reactions are k_1, k_2, and k_3, respectively.

This system of rate equations is known as the Lotka-Volterra equation and is most closely associated with population dynamics in predator-prey relationships. This system of equations has an oscillatory behavior. The amplitude of the oscillations depends on the concentration of A. Oscillations of this type are a form of emergent temporal order that is not present in equilibrium.

Real Examples

Real examples of clock reactions are the Belousov-Zhabotinsky reaction (BZ reaction), the Briggs-Rauscher reaction, the Bray-Liebhafsky reaction and the iodine clock reaction. These are oscillatory reactions, and the concentration of products and reactants can be approximated in terms of damped oscillations.

The best-known reaction, the BZ reaction, can be created with a mixture of potassium bromate ($KBrO_3$), malonic acid [$CH_2(COOH)_2$], and manganese sulfate ($MnSO_4$) prepared in a heated solution of sulfuric acid (H_2SO_4).

Spatial Order

An idealized example of spatial spontaneous symmetry breaking is the case in which we have two boxes of material separated by a permeable membrane so that material can diffuse between the two boxes. It is assumed that identical Brusselators are in each box with nearly identical initial conditions.

$$\frac{d}{dt}\{X_1\} = A + \{X_1\}^2\{Y_1\} - \{B\}\{X_1\} - \{X_1\} + D_x(X_2 - X_1)$$

$$\frac{d}{dt}\{Y_1\} = B\{X_1\} - \{X_1\}^2\{Y_1\} + D_y(Y_2 - Y_1)$$

$$\frac{d}{dt}\{X_2\} = \{A\} + \{X_2\}^2\{Y_2\} - \{B\}\{X_2\} - \{X_2\} + D_x(X_1 - X_2)$$

$$\frac{d}{dt}\{Y_2\} = B\{X_2\} - \{X_2\}^2\{Y_2\} + D_y(Y_1 - Y_2)$$

Here, the numerical subscripts indicate which box the material is in. There are additional terms proportional to the diffusion coefficient D that account for the exchange of material between boxes.

If the system is initiated with the same conditions in each box, then a small fluctuation will lead to separation of materials between the two boxes. One box will have a predominance of X, and the other will have a predominance of Y.

Biological Example

It is known that an important metabolic cycle, glycolysis, displays temporal order. Glycolysis consists of the degradation of one molecule of glucose and the overall production of two molecules of ATP. The process is therefore of great importance to the energetics of living cells. The global glycolysis reaction involves glucose, ADP, NAD, pyruvate, ATP, and NADH.

glucose + 2ADP + 2P_i + 2NAD$\rightarrow$ 2(pyruvate) + 2ATP + 2NADH

The details of the process are quite involved, however, a section of the process is autocatalyzed by Phosphofructokinase (PFK). This portion of the process is responsible for oscillations in the pathway that leads to the ability of the process to oscillate between an active and an inactive form. In other words, the autocatalytic reactions can turn the process off and on.

Phase Transitions

The initial amounts of reactants determine the distance from chemical equilibrium of the system. The greater the initial concentrations the further the system is from equilibrium. As the initial concentration increases, an abrupt change in order occurs. This abrupt change is known as phase transition. At the phase transition fluctuations in macroscopic quantities, such as chemical concentrations, increase as the system oscillates between the orderly state and the disorderly state. Also, at the phase transition, macroscopic equations, such as the rate equations, fail. Rate equations can be derived from microscopic considerations. The derivations typically rely on a mean field theory approximation to microscopic dynamical equations. Mean field theory breaks down in the presence of large fluctuations. Therefore, since large fluctuations occur in the neighborhood of a phase transition, macroscopic equations, such as rate equations, fail. As the initial concentration increases further, the system settle into an ordered state in which fluctuations are again small.

11

Pollution Prevention

Pollution prevention (P_2) describes activities that reduce the amount of pollution generated by a process, whether it is consumer consumption, driving, or industrial production. In contrast to most pollution control strategies, which seek to manage a pollutant after it is formed and reduce its impact upon the environment, the pollution prevention approach seeks to increase the efficiency of a process, thereby reducing the amount of pollution generated at its source. Although there is wide agreement that source reduction is the preferred strategy, some professionals also use the term pollution prevention to include recycling or reuse.

As an environmental management strategy, pollution prevention shares many attributes with cleaner production, a term used more commonly outside the United States. Pollution prevention encompasses more specialized sub-disciplines including green chemistry and green design (also known as environmentally conscious design).

The US Environmental Protection Agency has a number of P_2 programs that can assist individuals and organizations to implement P_2.

Source Reduction

Source reduction refers to any change in the design, manufacture, purchase, or use of materials or products (including packaging) to reduce their amount or toxicity before they become municipal solid waste.

Context

Source reduction is the highest goal in the waste management hierarchy - Reduce, Reuse, Recycle. It is a subset of Waste Reduction, which includes recycling and other strategies that decrease disposal quantities and pollution risk. Addressing polluting materials use as products are designed, produced, or used is the most dependable way to prevent waste from occurring.

Source Reduction is a design strategy – the creative application of Life Cycle Assessment and Alternatives Assessment and the preferred means to reach Zero waste targets and effect sustainable design. Source Reduction is important to Industrial Ecology as key to applying materials mapping and characterization results to practical and effective materials selection, placement, linking and usage.

Source reduction is the strategy of any individual, manufacturer, business, or other entity to eliminate or reduce the amount or toxicity of materials consumed, often reducing costs and other resources at the same time.

Source reduction is a set of tools and methods. A primary concept is the identification and characterization of problem materials to determine if safe use is a possibility or if safer alternatives will better fit in systems that support.

Synonyms

Pollution Prevention (or P2) and Toxics use reduction are also called source reduction because they address the use of hazardous substances at the source.

Procedures

Source Reduction is achieved through improvements in production and product design, or through Environmentally Preferable Purchasing (EPP).

Source Reduction in the United

In the United States, the Federal Trade Commission offers guidance for labelling claims: "Source reduction" refers to reducing or lowering the weight, volume or toxicity of a product or package. To avoid being misleading, source reduction claims must qualify the amount of the source reduction and give the basis for any comparison that is made. These principles apply regardless of whether a term like "source reduced" is used.

The Massachusetts Toxics Use Reduction Programme (TURA) offers 6 strategies to achieve source reduction:

- Toxic chemical substitution
- Production process modification
- Finished product reformulation
- Production modernization
- Improvements in operations and maintenance
- In-process recycling of production material.

Environmentalism

Environmentalism is a broad philosophy and social movement centered on a concern for the conservation and improvement of the environment. Environmentalism is associated with the color green.

Environmentalism as a Social Movement

Environmentalism can also be defined as a social movement which seeks to influence the political process by lobbying, activism, and education in order to protect natural resources and ecosystems. In recognition of humanity as a participant in ecosystems, the environmental movement is centered on ecology, health, and human rights.

An *environmentalist* is a person who may advocate the sustainable management of resources and stewardship of the natural environment through changes in public policy or individual behavior. In various ways (for example, grassroots activism and protests), environmentalists and environmental organizations seek to give the natural world a stronger voice in human affairs. Though opinions vary, environmentalism may be seen as a spectrum; from the reformist to the radical.

History

This section needs additional citations for verification. Please help improve this article by adding reliable references (ideally, using *inline citations*). Unsourced material may be challenged and removed. Though the modern environmental movement arose during the Industrial Revolution, a concern for environmental protection has recurred in diverse forms, in different parts of the world, throughout history. For example, in the Middle East, the earliest known writings concerned with environmental pollution were Arabic medical treatises written during the "Arab Agricultural Revolution", by writers such as Alkindus, Costa ben Luca, Rhazes, Ibn Al-Jazzar, al-Tamimi, al-Masihi, Avicenna, Ali ibn Ridwan, Isaac Israeli ben Solomon, Abd-el-latif, and Ibn al-Nafis. They were concerned with air contamination, water contamination, soil contamination, solid waste mishandling, and environmental assessments of certain localities

In Europe, King Edward I of England banned the burning of sea-coal by proclamation in London in 1272, after its smoke had become a problem. But the fuel was so common in England that this earliest of names for it was acquired because it could be carted away from some shores by the wheelbarrow. Air pollution would continue to be a problem there, especially later during the industrial revolution, and extending into the recent past with the Great Smog of 1952.

Origins of Environmental Movement

In Europe, it was the Industrial Revolution that gave rise to modern environmental pollution as it is generally understood today. The emergence of great factories and consumption of immense quantities of coal and other fossil fuels gave rise to unprecedented air pollution and the large volume of industrial chemical discharges added to the growing load of untreated human waste. The first large-scale, modern environmental laws came in the form of the British Alkali Acts, passed in 1863, to regulate the deleterious air pollution (gaseous hydrochloric acid) given off by the Leblanc process, used to produce soda ash. Environmentalism grew out of the amenity movement, which was a reaction to industrialization, the growth of cities, and worsening air and water pollution.

In the United States, the beginnings of an environmental movement can be traced as far back as 1739, when Benjamin Franklin and other Philadelphia residents, citing "public rights," petitioned the Pennsylvania Assembly to stop waste dumping and remove tanneries from Philadelphia's commercial district. The US movement expanded in the 1800s, out of concerns for protecting the natural resources of the West, with individuals such as John Muir and Henry David Thoreau making key philosophical contributions. Thoreau was interested in peoples' relationship with nature and studied this by living close to nature in a simple life. He published his experiences in the book *Walden,* which argues that people should become intimately close with nature. Muir came to believe in nature's inherent right, especially after spending time hiking in Yosemite Valley and studying both the ecology and geology. He successfully lobbied congress to form Yosemite National Park and went on to set up the Sierra Club. The conservationist principles as well as the belief in an inherent right of nature were to become the bedrock of modern environmentalism.

In the 20th century, environmental ideas continued to grow in popularity and recognition. Efforts were starting to be made to save some wildlife, particularly the American Bison. The death of the last Passenger Pigeon as well as the endangerment of the American Bison helped to focus the minds of conservationists and popularize their concerns. Notably in 1916 the National Park Service was founded by President Woodrow Wilson.

In 1949, *A Sand County Almanac* by Aldo Leopold was published. It explained Leopold's belief that humankind should have moral respect for the environment and that it is unethical to harm it. The book is sometimes called the most influential book on conservation.

In 1962, Houghton Mifflin published *Silent Spring* by American biologist Rachel Carson. The book cataloged the environmental impacts of the indiscriminate spraying of DDT in the US and questioned the logic of releasing large amounts of chemicals into the environment without fully understanding their effects on ecology or human health. The book suggested that DDT and other pesticides may cause cancer and that their agricultural use was a threat to wildlife, particularly birds. The resulting public concern lead to the creation of the United States Environmental Protection Agency in 1970 which subsequently banned the agricultural use of DDT in the US in 1972. The limited use of DDT in disease vector control continues to this day in certain parts of the world and remains controversial. The book's legacy was to produce a far greater awareness of environmental issues and interest into how people affect the environment. With this new interest in environment came interest in problems such as air pollution and oil spills, and environmental interest grew. New pressure groups formed, notably Greenpeace and Friends of the Earth.

In the 1970s, the Chipko movement was formed in India; influenced by Mahatma Gandhi, they set up peaceful resistance to deforestation by literally hugging trees (leading to the term "tree huggers"). Their peaceful methods of protest and slogan "ecology is permanent economy" were very influential.

By the mid-1970s, many felt that people were on the edge of environmental catastrophe. The Back-to-the-land movement started to form and ideas of environmental ethics joined with anti-Vietnam War sentiments and other political issues. These individuals lived outside normal society and started to take on some of the more radical environmental theories such as deep ecology. Around this time more mainstream environmentalism was starting to show force with the signing of the Endangered Species Act in 1973 and the formation of CITES in 1975.

In 1979, James Lovelock, a former NASA scientist, published *Gaia: A new look at life on Earth*, which put forth the Gaia Hypothesis; it proposes that life on Earth can be understood as a single organism. This became an important part of the Deep Green ideology. Throughout the rest of the history of environmentalism there has been debate and argument between more radical followers of this Deep Green ideology and more mainstream environmentalists.

Environmentalism has also changed to deal with new issues such as global warming and genetic engineering.

Environmental Movement

The environmental movement (a term that sometimes includes the conservation and green movements) is a diverse scientific, social, and political movement. In general terms, environmentalists advocate the sustainable management of resources, and the protection (and restoration, when necessary) of the natural environment through changes in public policy and individual behavior. In its recognition of humanity as a participant in ecosystems, the movement is centered around ecology, health, and human rights.

Additionally, throughout history, the movement has been incorporated into religion. The movement is represented by a range of organizations, from the large to grassroots, but a younger demographic than is common in other social movements. Due to its large membership, varying and strong beliefs, the movement is not entirely united. Indeed, some argue that an environmental ethic of at least some sort is so urgently needed in all quarters that the broader the better. Conversely, disunity can be a weakness in the face of strong opposition from unsympathetic political and industrial forces.

Free Market Environmentalism

Free market environmentalism is a theory that argues that the free market, property rights, and tort law provide the best tools to preserve the health and sustainability of the environment. This is in sharp contrast to the most common approach of looking to legislative government intervention to prevent destruction of the environment. It considers environmental stewardship to be natural, as well as the expulsion of pollutors and other aggressors through individual and class action.

Environmental preservation, chiefly in the United States, is viewed as the strict setting aside of natural resources to prevent damage caused by contact with humans or by certain human activities, such as logging, mining, hunting, and fishing. It is different from conservation; conservation allows for some degree of industrial development, albeit it within sustainable limits Regulations and laws may be enacted for the preservation of natural resources.

Elsewhere in the world the terms preservation and conservation may be less contested and are often used interchangeably.

Dark Greens, Light Greens and Bright Greens

Contemporary environmentalists are often described as being split into three groups:

- Dark;
- Light; and
- Bright Greens.

Light Greens see protecting the environment first and foremost as a personal responsibility. They fall in on the reformist end of the spectrum introduced above, but light Greens do not emphasize environmentalism as a distinct political ideology, or even seek fundamental political reform. Instead they often focus on environmentalism as a lifestyle choice The motto "Green is the new black." sums up this way of thinking, for many.

In contrast, Dark Greens believe that environmental problems are an inherent part of industrialized capitalism, and seek radical political change. As discussed earlier, 'dark greens' tend to believe that dominant political ideologies (sometimes referred to as industrialism) are corrupt and inevitably lead to consumerism, alienation from nature and resource depletion. Dark Greens claim that this is caused by the emphasis on growth that exists within all existing ideologies, a tendency referred to as 'growth mania'. The dark green brand of environmentalism is associated with ideas of Deep Ecology, Post-materialism, Holism, the Gaia Theory of James Lovelock and the work of Fritjof Capra. The division between light and dark greens was visible in the fighting between Fundi and Realo factions of the German Green Party Since the Dark Greens often embrace strands of communist and marxist philosophies, the motto "Green is the new red." is often used in describing their beliefs.

More recently, a third group may be said to have emerged in the form of Bright Greens. This group believes that radical changes are needed in the economic and political operation of society in order to make it sustainable, but that better designs, new technologies and more widely distributed social innovations are the means to make those changes— and that we can neither shop nor protest our way to sustainability. As Ross Robertson writes, "[B]right green environmentalism is less about the problems and limitations we need to overcome than the "tools, models, and ideas" that already exist for overcoming them. It forgoes the bleakness of protest and dissent for the energizing confidence of constructive solutions."

Environmental Organizations and Conferences

Environmental Organizations can be global, regional, national or local; they can be government-run or private (NGO). Despite a tendency to see environmentalism as an American or Western-centered pursuit, almost every country has its share of environmental activism. Moreover, groups dedicated to community development and social justice may also attend to environmental concerns.

There are some volunteer organizations. For example Ecoworld, which is about the environment and is based in team work and volunteer work. Some US environmental organizations, among them the Natural Resources Defense Council and the Environmental Defense Fund, specialize in bringing lawsuits (a tactic seen as particularly useful in that country). Other groups, such as the US-based National Wildlife Federation, the Nature Conservancy, and The Wilderness Society, and global groups like the World Wide Fund (WWF) for Nature and Friends of the Earth, disseminate information, participate in public hearings, lobby, stage demonstrations, and may purchase land for preservation. Smaller groups, including Wildlife Conservation International, conduct research on endangered species and ecosystems. More radical organizations, such as Greenpeace, Earth First!, and the Earth Liberation Front, have more directly opposed actions they regard as environmentally harmful. While Greenpeace is devoted to nonviolent confrontation as a means of bearing witness to environmental wrongs and bringing issues into the public realm for debate, the underground Earth Liberation Front engages in the clandestine destruction of property, the release of caged or penned animals, and other criminal acts. Such tactics are regarded as unusual within the movement, however.

On an international level, concern for the environment was the subject of a UN conference in Stockholm in 1972, attended by 114 nations. Out of this meeting developed UNEP (United Nations Environment Programme) and the follow-up United Nations Conference on Environment and Development in 1992. Other international organizations in support of environmental policies development include the Commission for Environmental Cooperation (NAFTA), the European Environment Agency (EEA), and the Inter-governmental Panel on Climate Change (IPCC).

12

Energy Conservation

Energy conservation is the practice of decreasing the quantity of energy used. It may be achieved through efficient energy use, in which case energy use is decreased while achieving a similar outcome, or by reduced consumption of energy services. Energy conservation may result in increase of financial capital, environmental value, national security, personal security, and human comfort. Individuals and organizations that are direct consumers of energy may want to conserve energy in order to reduce energy costs and promote economic security. Industrial and commercial users may want to increase efficiency and thus maximize profit.

Electrical energy conservation is an important element of energy policy. Energy conservation reduces the energy consumption and energy demand per capita and thus offsets some of the growth in energy supply needed to keep up with population growth. This reduces the rise in energy costs, and can reduce the need for new power plants, and energy imports. The reduced energy demand can provide more flexibility in choosing the most preferred methods of energy production.

By reducing emissions, energy conservation is an important part of lessening climate change. Energy conservation facilitates the replacement of non-renewable resources with renewable energy. Energy conservation is often the most economical solution to energy shortages, and is a more environmentally benign alternative to increased energy production.

United States

The United States is currently the largest single consumer of energy. The U.S. Department of Energy categorizes national energy use in four broad sectors: transportation, residential, commercial, and industria.

Energy usage in transportation and residential sectors (about half of U.S. energy consumption) is largely controlled by individual domestic consumers. Commercial and industrial energy expenditures are determined by businesses entities and other facility managers. National energy policy has a significant effect on energy usage across all four sectors.

Transportation

The transportation includes all vehicles used for personal or freight transportation. Of the energy used in this sector, approximately 65% is consumed by gasoline-powered vehicles, primarily personally owned. Diesel-powered transport (trains, merchant ships, heavy trucks, etc.) consumes about 20%, and air traffic consumes most of the remaining 15%.

The two oil supply crisis of the 1970s spurred the creation, in 1975, of the federal Corporate Average Fuel Economy (CAFE) program, which required auto manufacturers to meet progressively higher fleet fuel economy targets. The next decade saw dramatic improvements in fuel economy, mostly the result of reductions in vehicle size and weight which originated in the late 1970s, along with the transition to front wheel drive. These gains eroded somewhat after 1990 due to the growing popularity of sport utility vehicles, pickup trucks and minivans, which fall under the more lenient "light truck" CAFE standard.

In addition to the CAFE program, the U.S. government has tried to encourage better vehicle efficiency through tax policy. Since 2002, taxpayers have been eligible for income tax credits for gas/electric hybrid vehicles. A "gas-guzzler" tax has been assessed on manufacturers since 1978 for cars with exceptionally poor fuel economy. While this tax remains in effect, it currently generates very little revenue as overall fuel economy has improved. The gas-guzzler tax ended the reign of large cubic-inched engines from the musclecar era.

Another focus in gasoline conservation is reducing the number of miles driven. An estimated 40% of American automobile use is associated with daily commuting. Many urban areas offer subsidized public transportation to reduce commuting traffic, and encourage carpooling by providing designated high-occupancy vehicle lanes and lower tolls for cars with multiple riders. In recent years telecommuting has also become a viable alternative to commuting for some jobs, but in 2003 only 3.5% of workers were telecommuters. Ironically, hundreds of thousands of American and European workers have been replaced by workers in Asia who telecommute from thousands of miles away.

Fuel economy-maximizing behaviors also help reduce fuel consumption. Among the most effective are moderate (as opposed to aggressive) driving, driving at lower speeds, using cruise control, and turning off a vehicle's engine at stops rather than idling. A vehicle's gas mileage decreases rapidly highway speeds, normally above 55 miles per hour (though the exact number varies by vehicle). This is because aerodynamic forces are proportionally related to the square of an object's speed (when the speed is doubled, drag quadruples). According to the U.S. Department of Energy (DOE), as a rule of thumb, each 5 mph (8.0 km/h) you drive over 60 mph (97 km/h) is similar to paying an additional $0.30 per gallon for gas. The exact speed at which a vehicle achieves it's highest efficiency varies based on the vehicle's drag coefficient, frontal area, surrounding air speed, and the efficiency and gearing of a vehicle's drive train and transmission.

Residential Sector

The residential sector refers to all private residences, including single-family homes, apartments, manufactured homes and dormitories. Energy use in this sector varies significantly across the country, due to regional climate differences and different regulation. On average, about half of the energy used in U.S. homes is expended on space conditioning (i.e. heating and cooling).

The efficiency of furnaces and air conditioners has increased steadily since the energy crises of the 1970s. The 1987 National Appliance Energy Conservation Act authorized the Department of Energy to set minimum efficiency standards for space conditioning equipment and other appliances each year, based on what is "technologically feasible and economically justified". Beyond these minimum standards, the Environmental Protection Agency awards the Energy Star designation to appliances that exceed industry efficiency averages by an EPA-specified percentage.

Despite technological improvements, many American lifestyle changes have put higher demands on heating and cooling resources. The average size of homes built in the United States has increased significantly, from 1,500 sq ft (140 m^2) in 1970 to 2,300 sq ft (210 m^2) in 2005. The single-person household has become more common, as has central air conditioning: 23% of households had central air conditioning in 1978, that figure rose to 55% by 2001.

As furnace efficiency gets higher, there is limited room for improvement—efficiencies above 85% are now common. However, improving the building envelope through better or more insulation, advanced windows, etc., can allow larger improvements. The passive house approach produces superinsulated buildings that approach zero net energy consumption. Improving the building envelope can also be cheaper than replacing a furnace or air conditioner.

Even lower cost improvements include weatherization, which is frequently subsidized by utilities or state/federal tax credits, as are programmable thermostats. Consumers have also been urged to adopt a wider indoor temperature range (e.g. 65 °F (18 °C) in the winter, 80 °F (27 °C) in the summer).

One underutilized, but potentially very powerful means to reduce household energy consumption is to provide real-time feedback to homeowners so they can effectively alter their energy using behavior. Recently, low cost energy feedback displays, such as The Energy Detective or wattson have become available. A study of a similar device deployed in 500 Ontario homes by *Hydro One* showed an average 6.5% drop in total electricity use when compared with a similarly sized control group.

Standby power used by consumer electronics and appliances while they are turned off accounts for an estimated 5 to 10% of household electricity consumption, adding an estimated $3 billion to annual energy costs in the USA. "In the average home, 75% of the electricity used to power home electronics is consumed while the products are turned off."

Energy usage in some homes may vary widely from these averages. For example, milder regions such as the southern U.S. and Pacific coast of the USA need far less energy for space conditioning than New York City or Chicago. On the other hand, air conditioning energy use can be quite high in hot-arid regions (Southwest) and hot-humid zones (Southeast) In milder climates such as San Diego, lighting energy may easily consume up to 40% of total energy. Certain appliances such as a waterbed, hot tub, or pre-1990 refrigerator use significant amounts of electricity. However, recent trends in home entertainment equipment can make a large difference in household energy use. For instance a 50" LCD television (average on-time= 6 hours a day) may draw 300 Watts less than a similarly sized plasma system. In most residences no single appliance dominates, and any conservation efforts must be directed to numerous areas in order to achieve substantial energy savings. However, Ground,

Air and Water Source Heat Pump systems are the more energy efficient, environmentally clean, and cost-effective space conditioning and domestic hot water systems available (Environmental Protection Agency), and can achieve reductions in energy consumptions of up to 69 per cent.

Best Building Practices

Current best practices in building design, construction and retrofitting result in homes that are profoundly more energy conserving than average new homes. This includes insulation and energy-efficient windows and lighting See Passive house, Superinsulation, Self-sufficient homes, Zero energy building, Earthship, MIT Design Advisor, Energy Conservation Code for Indian Commercial Buildings.

Smart ways to construct homes such that minimal resources are used to cooling and heating the house in summer and winter respectively can significantly reduce energy costs.

Commercial Sector

The commercial sector consists of retail stores, offices (business and government), restaurants, schools and other workplaces. Energy in this sector has the same basic end uses as the residential sector, in slightly different proportions. Space conditioning is again the single biggest consumption area, but it represents only about 30% of the energy use of commercial buildings. Lighting, at 25%, plays a much larger role than it does in the residential sector Lighting is also generally the most wasteful component of commercial use. A number of case studies indicate that more efficient lighting and elimination of over-illumination can reduce lighting energy by approximately fifty percent in many commercial buildings.

Commercial buildings can greatly increase energy efficiency by thoughtful design, with today's building stock being very poor examples of the potential of systematic (not expensive) energy efficient design. Commercial buildings often have professional management, allowing centralized control and coordination of energy conservation efforts. As a result, fluorescent lighting (about four times as efficient as incandescent) is the standard for most commercial space, although it may produce certain adverse health effects Potential health concerns can be mitigated by using newer fixtures with electronic ballasts rather than older magenetic ballasts. As most buildings have consistent hours of operation, programmed thermostats and lighting controls are common. However, too many companies believe that merely having a computer controlled Building automation system guarantees energy efficiency. As an example one large company in Northern California boasted that it was confident its state of the art system had optimized space heating. A more careful analysis by Lumina Technologies showed the system had been given programming instructions to maintain constant 24 hour temperatures in the entire building complex. This instruction caused the injection of nighttime heat into vacant buildings when the daytime summer temperatures would often exceed 90 °F (32 °C). This mis-programming was costing the company over $130,000 per year in wasted energy (Lumina Technologies, 1997). Many corporations and governments also require the Energy Star rating for any new equipment purchased for their buildings.

Solar heat loading through standard window designs usually leads to high demand for air conditioning in summer months. An example of building design overcoming this excessive heat loading is the Dakin Building in Brisbane, California, where fenestration was designed to achieve an angle with respect to sun incidence to allow maximum reflection of solar heat; this design also assisted in reducing interior over-illumination to enhance worker efficiency and comfort.

Advances include use of occupancy sensors to turn off lights when spaces are unoccupied, and photosensors to dim or turn off electric lighting when natural light is available. In air conditioning systems, overall equipment efficiencies have increased as energy codes and consumer information have begun to emphasise year round performance rather than just efficiency ratings at maximum output. Controllers that automatically vary the speeds of fans, pumps, and compressors have radically improved part-load performance of those devices. For space or water heating, electric heat pumps consume roughly half the energy required by electric resistance heaters. Natural gas heating efficiencies have improved through use of condensing furnaces and boilers, in which the water vapor in the flue gas is cooled to liquid form before it is discharged, allowing the heat of condensation to be used. In buildings where high levels of outside air are required, heat exchangers can capture heat from the exhaust air to preheat incoming supply air.

Industrial Sector

The industrial sector represents all production and processing of goods, including manufacturing, construction, farming, water management and mining. Increasing costs have forced energy-intensive industries to make substantial efficiency improvements in the past 30 years. For example, the energy used to produce steel and paper products has been cut 40% in that time frame, while petroleum/aluminum refining and cement production have reduced their usage by about 25%. These reductions are largely the result of recycling waste material and the use of cogeneration equipment for electricity and heating.

Another example for efficiency improvements is the use of products made of High temperature insulation wool (HTIW) which enables predominantly industrial users to operate thermal treatment plants at temperatures between 800 and 1400 °C. In these high-temperature applications, the consumption of primary energy and the associated CO_2 emissions can be reduced by up to 50% compared with old fashioned industrial installations. The application of products made of High temperature insulation Wool is becoming increasingly important against the background of the currently dramatic rising cost of energy.

The energy required for delivery and treatment of fresh water often constitutes a significant percentage of a region's electricity and natural gas usage (an estimated 20% of California's total energy use is water-related. In light of this, some local governments have worked toward a more integrated approach to energy and water conservation efforts.

To conserve energy, some industries have begun using solar panels to heat their water.

Unlike the other sectors, total energy use in the industrial sector has declined in the last decade. While this is partly due to conservation efforts, it's also a reflection of the growing trend for U.S. companies to move manufacturing operations overseas.

United Kingdom

Energy conservation in the United Kingdom has been receiving increased attention over recent years. Key factors behind this are the Government's commitment to reducing carbon emissions, the projected 'energy gap' in UK electricity generation, and the increasing reliance on imports to meet national energy needs. Domestic housing and road transport are currently the two biggest problem areas.

The UK Government has jointly funded the Energy Saving Trust to promote energy conservation at a consumer, business and community level since 1993.

Issues with Energy Conservation

Critics and advocates of some forms of energy conservation make the following arguments:

- Standard economic theory suggests that technological improvements that increase energy efficiency will tend to increase, rather than reduce energy use. This is called the *Jevons Paradox* and it is said to occur in two ways. Firstly, increased energy efficiency makes the use of energy relatively cheaper, thus encouraging increased use. Secondly, increased energy efficiency leads to increased economic growth, which pulls up energy use in the whole economy. This does not imply that increased fuel efficiency is worthless. Increased fuel efficiency enables greater production and a higher quality of life.

- Some retailers argue that bright lighting stimulates purchasing. Health studies have demonstrated that headache, stress, blood pressure, fatigue and worker error all generally increase with the common over-illumination present in many workplace and retail settings. It has been shown that natural daylighting increases productivity levels of workers, while reducing energy consumption.
- The use of telecommuting by major corporations is a significant opportunity to conserve energy, as many Americans now work in service jobs that enable them to work from home instead of commuting to work each day.
- Electric motors consume more than 60% of all electrical energy generated and are responsible for the loss of 10 to 20% of all electricity converted into mechanical energy.
- Consumers are often poorly informed of the savings of energy efficient products. The research one must put into conserving energy often is too time consuming and costly when there are cheaper products and technology available using today's fossil fuels.

Efficient energy use, sometimes simply called energy efficiency, is using less energy to provide the same level of energy service. An example would be insulating a home to use less heating and cooling energy to achieve the same temperature. Another example would be installing fluorescent lights and/or skylights instead of incandescent lights to attain the same level of illumination. Efficient energy use is achieved primarily by means of a more efficient technology or process rather than by changes in individual behaviour.

Energy efficient buildings, industrial processes and transportation could reduce the world's energy needs in 2050 by one third, and help controlling global emissions of greenhouse gases, according to the International Energy Agency.

Energy efficiency and renewable energy are said to be the "twin pillars" of sustainable energy policy.

Overview

Making homes, vehicles, and businesses more energy efficient is seen as a largely untapped solution to addressing global warming, energy security, and fossil fuel depletion. Many of these ideas have been discussed for years, since the 1973 oil crisis brought energy issues to the forefront. In the late 1970s, physicist Amory Lovins popularized the notion of a "soft energy path", with a strong focus on energy efficiency. Among other things, Lovins popularized the notion of negawatts—the idea of meeting energy needs by increasing efficiency instead of increasing energy production.

Energy efficiency has proved to be a cost-effective strategy for building economies without necessarily growing energy consumption, as environmental business strategist Joel Makower has noted. For example, the state of California began implementing energy-efficiency measures in the mid-1970s, including building code and appliance standards with strict efficiency requirements. During the following years, California's energy consumption has remained approximately flat on a per capita basis while national U.S. consumption doubled. As part of its strategy, California implemented a three-step plan for new energy resources that puts energy efficiency first, renewable electricity supplies second, and new fossil-fired power plants last.

Still, efficiency often has taken a secondary position to new power generation as a solution to global warming in creating national energy policy. Some companies also have been reluctant to engage in efficiency measures, despite the often favorable returns on investments that can result. Lovins' Rocky Mountain Institute points out that in industrial settings, "there are abundant opportunities to save 70% to 90% of the energy and cost for lighting, fan, and pump systems; 50% for electric motors; and 60% in areas such as heating, cooling, office equipment, and appliances." In general, up to 75% of the electricity used in the U.S. today could be saved with efficiency measures that cost less than the electricity itself.

Other studies have emphasized this. A report published in 2006 by the McKinsey Global Institute, asserted that "there are sufficient economically viable opportunities for energy-productivity improvements that could keep global energy-demand growth at less than one per cent per annum" — less than half of the 2.2 per cent average growth anticipated through 2020 in a business-as-usual scenario. Energy productivity — which measures the output and quality of goods and services per unit of energy input — can come from either reducing the amount of energy required to produce something, or from increasing the quantity or quality of goods and services from the same amount of energy.

The Vienna Climate Change Talks 2007 Report, under the auspices of the United Nations Framework Convention on Climate Change (UNFCCC), clearly shows "that energy efficiency can achieve real emission reductions at low cost."

Energy Efficient Appliances

Modern energy-efficient appliances, such as refrigerators, freezers, ovens, stoves, dishwashers, and clothes washers and dryers, use significantly less energy than older appliances. Current energy efficient refrigerators, for example, use 40 percent less energy than conventional models did in 2001. Modern power management systems also reduce energy usage by idle appliances by turning them off or putting them into a low-energy mode after a certain time. Many countries identify energy-efficient appliances using an Energy Star label.

Energy Efficient Building Design

A building's location and surroundings play a key role in regulating its temperature and illumination. For example, trees, landscaping, and hills can provide shade and block wind. In cooler climates, designing buildings with an east-west orientation to increase the number of south-facing windows minimizes energy use, by maximizing passive solar heating. Tight building design, including energy-efficient windows, well-sealed doors, and additional thermal insulation of walls, basement slabs, and foundations can reduce heat loss by 25 to 50 per cent

Dark roofs may become up to 70 °F hotter than the most reflective white surfaces, and they transmit some of this additional heat inside the building. US Studies have shown that lightly colored roofs use 40 per cent less energy for cooling than buildings with darker roofs. White roof systems save more energy in sunnier climates. Advanced electronic heating and cooling systems can moderate energy consumption and improve the comfort of people in the building.

Proper placement of windows and skylights and use of architectural features that reflect light into a building, can reduce the need for artificial lighting. Compact fluorescent lights use two-thirds less energy and may last 6 to 10 times longer than incandescent light bulbs. Newer fluorescent lights produce a natural light, and in most applications they are cost effective, despite their higher initial cost, with payback periods as low as a few months However, those ideals may not always be achieved in practice, because lifetime depends on the frequency of usage. In addition, CFLs emit UV light which can harm paintings, textiles and pigments. They also respond more slowly when switched on, so may represent a safety hazard in halls and stairways for example. While incandescent bulbs do contribute to the space heating of a building, the heat that they produce, being electrically produced, is probably more expensive and certainly more carbon-intensive than, for example, gas-fired heating. Furthermore, any heat that such bulbs produce during the summer is likely to be unwanted and may lead to yet more electrical demand for space cooling. Increased use of natural and task lighting have been shown by one study to increase productivity in schools and offices However, fluorescent lighting can be harsh, and the flicker can induce migraine, so caution is needed when replacing incandescent lights. Modern compact fluorescent lighting can produce a warmer and less harsh light.

Effecive energy-efficient building design can include the use of low cost Passive Infra Reds (PIRs) to switch-off lighting when areas are unnoccupied such as toilets, corridors or even office areas out-of-hours. In addition, lux levels can be monitored using daylight sensors linked to the building's lighting scheme to switch on/off or dim the lighting to pre-defined levels to take into account the natural light and thus reduce consumption. Building Management Systems (BMS) link all of this together in one centralised computer to control the whole building's lighting and power requirements.

Smart meters are slowly being adopted by the commerial sector to highlight to staff and for internal monitoring purposes the building's energy usage in a dynamic presentable format. The use of Power Quality Analysers can be introduced into an existing building to assess usage, harmonic distortion, peaks, swells and interruptions amongst others to ultimately make the building more energy-efficient.

Energy Efficiency for Industry

In industry, when electricity is generated, the heat which is produced as a by-product can be captured and used for process steam, heating or other industrial purposes. Conventional electricity generation is about 30 percent efficient, whereas combined heat and power (also called cogeneration) converts up to 90 per cent of the fuel into usable energy.

Advanced boilers and furnaces can operate at higher temperatures while burning less fuel. These technologies are more efficient and produce fewer pollutants.

Over 45 per cent of the fuel used by US manufacturers is burnt to make steam. The typical industrial facility can reduce this energy usage 20 percent (according to the US Department of Energy) by insulating steam and condensate return lines, stopping steam leakage, and maintaining steam traps.

Electric motors usually run on a constant flow of energy, but an adjustable speed drive can vary the motor's energy output to match the load. This achieves energy savings ranging from 3 to 60 percent, depending on how the motor is used. Motor coils made of superconducting materials can also reduce energy losses Motors may also benefit from voltage optimisation.

Many industries use compressed air for sand blasting, painting, or other tools. According to the US Department of Energy, optimizing compressed air systems by installing variable speed drives, along with preventive maintenance to detect and fix air leaks, can improve energy efficiency 20 to 50 per cent.

Energy Efficient Vehicles

Further information: Automotive market and Alternative propulsion. Using improved aerodynamics to minimize drag can increase vehicle fuel efficiency. Reducing vehicle weight can significantly also improve fuel economy, which is why composite materials are widely used in car bodies.

More advanced tires, with decreased tire to road friction and rolling resistance, can save gasoline. Fuel economy can be improved over three percent by keeping tires inflated to the correct pressure. Replacing a clogged air filter can improve a cars fuel consumption by as much as 10 per cent.

Fuel efficient vehicles may reach twice the fuel efficiency of the average automobile. Cutting-edge designs, such as the diesel Mercedes-Benz Bionic concept vehicle have achieved a fuel efficiency as high as 84 miles per US gallon (2.8 L/100 km; 101 mpg_{-imp}), four times the current conventional automotive average.

Another growing trend in automotive efficiency is the rise of hybrid and electric cars. Hybrids, like the Toyota Prius, use regenerative braking to recapture energy that would dissipate in normal cars; the effect is especially pronounced in city driving. plug-in hybrids also have electrical plugs, which makes it possible to drive for limited distances without burning any gasoline; in this case, energy efficiency is dictated by whatever process (coal-burning, hydroelectric, etc.) created the power. Plug-ins can typically drive for around 40 mile purely on electricity without recharging; if the battery runs low, a gas engine kicks in allowing for extended range. Finally, all-electric cars are also growing in popularity; the Tesla Roadster sports car is the only high-performance all-electric car currently on the market, and others are in design.

Energy Conservation

Energy conservation is broader than energy efficiency in that it encompasses using less energy to achieve a lesser energy service, for example through behavioural change, as well as encompassing energy efficiency. Examples of conservation without efficiency improvements would be heating a room less in winter, driving less, or working in a less brightly lit room. As with other definitions, the boundary between efficient energy use and energy conservation can be fuzzy, but both are important in environmental and economic terms. This is especially the case when actions are directed at the saving of fossil fuels.

Sustainable Energy

Energy efficiency and renewable energy are said to be the "twin pillars" of a sustainable energy policy. Both strategies must be developed concurrently in order to stabilize and reduce carbon dioxide emissions in our lifetimes. Efficient energy use is essential to slowing the energy demand growth so that rising clean energy supplies can make deep cuts in fossil fuel use. If energy use grows too rapidly, renewable energy development will chase a receding target. Likewise, unless clean energy supplies come online rapidly, slowing demand growth will only begin to reduce total carbon emissions; a reduction in the carbon content of energy sources is also needed. A sustainable energy economy thus requires major commitments to both efficiency and renewables.

Rebound Effect

If the demand for energy services remains constant, improving energy efficiency will reduce energy consumption and carbon emissions. However, many efficiency improvements do not reduce energy consumption by the amount predicted by simple engineering models. This is because they make energy services cheaper, and so consumption of those services increase. For example, since fuel efficient vehicles make travel cheaper, consumers may choose to drive further and/or faster, thereby offsetting some of the potential energy savings. This is an example of the direct *rebound effect.*

Estimates of the size of the rebound effect range from roughly 5% to 40%. Rebound effects are smaller in mature markets where demand is saturated. The rebound effect is likely to be less than 30% at the household level and may be closer to 10% for transport A rebound effect of 30% implies that improvements in energy efficiency should achieve 70% of the reduction in energy consumption projected using engineering models.

Since more efficient (and hence cheaper) energy will also lead to faster economic growth, there are suspicions that improvements in energy efficiency may eventually lead to even faster resource use. This was postulated by economists in the 1980's and remains a controversial hypothesis. Ecological economists have suggested that any cost savings from efficiency gains to save energy is a coalition consisting largely of industrial, technological, and energy corporations. The Alliance states that its mission is to "support energy efficiency as a cost-effective energy resource under existing market conditions and advocate energy-efficiency policies that minimize costs to society and individual consumers, and that lessen greenhouse gas emissions and their impact on the global climate." The Alliance's chief activities include public relations, research, and lobbying to change U.S. energy policy.

Conservation Movement

The conservation movement also known as nature conservation is a political, social and, to some extent, scientific movement that seeks to protect natural resources including plant and animal species as well as their habitat for the future.

The early conservation movement included fisheries and wildlife management, water, soil conservation and sustainable forestry. The contemporary conservation movement has broadened from the early movement's emphasis on use of sustainable yield of natural resources and preservation of wilderness areas to include preservation of biodiversity. Some say the conservation movement is part of the broader and more far-reaching environmental movement, while others argue that they differ both in ideology and practice. Chiefly in the United States, conservation is seen as differing from environmentalism in that it aims to preserve natural resources expressly for their continued sustainable use by humans. In other parts of the world conservation is used more broadly to include the setting aside of natural areas and the active protection of wildlife for their inherent value, as much as for any value they may have for humans.

Early Worldwide Conservation Movement

The nascent conservation movement slowly developed in the 19th century, starting first in the scientific forestry methods pioneered by the Germans and the French in the 17th and 18th centuries. While continental Europe created the scientific methods later used in conservationist efforts, British India and the United States are credited with starting the conservation movement.

Foresters in India, often German, managed forests using early climate change theories that Alexander von Humboldt developed in the mid 19th century, applied fire protection, and tried to keep the "house-hold" of nature. This was an early ecological idea, in order to preserve the growth of delicate teak trees. The same German foresters who headed the Forest Service of India, such as Dietrich Brandis and Berthold Ribbentrop, traveled back to Europe and taught at forestry schools in England (Cooper's Hill, later moved to Oxford). These men brought with them the legislative and scientific knowledge of conservationism in British India back to Europe, where they distributed it to men such as Gifford Pinchot, which in turn helped bring European and British Indian methods to the United States.

Philosophy of Early American Conservation Movement

During the nineteenth century, Americans developed a deep and abiding passion for nature. The early evolution of the conservation movement began through both public and private recognition of the relationship between man and nature often reflected in the great literary and artistic works of the nineteenth century Artists, such as Albert Beirstadt, painted powerful landscapes of the American West during the mid nineteenth century, which were incredibly popular images representative of the unique natural wonders of the American frontier Likewise, in 1860, Frederic Edwin Church painted "Twilight in the Wilderness", which was an artistic masterpiece of the era that explored the growing importance of the American wilderness.

Many American writers also romanticized and focused upon nature as a subject matter. However, the most notable literary figure upon the early conservation movement proved to be Henry David Thoreau. Throughout his work, *Walden,* Thoreau detailed his experiences at the natural setting of Walden Pond and his deep appreciation for nature. In one instance, he described a deep grief for a tree that was cut down. Thoreau went on to bemoan the lack of reverence for the natural world: "I would that our farmers when they cut down a forest felt some of that awe which the old Romans did when they came to thin, or let in the light to, a consecrated grove". As he states in *Walden,* Thoreau "was interested in the preservation" of nature In 1860, Henry David Thoreau delivered a speech to the Middlesex Agricultural Society in Massachusetts; the speech, entitled "The Succession of Forest Trees", explored forest ecology and encouraged the agricultural community to plant trees. This speech became one of Thoreau's "most influential ecological contributions to conservationist thought".

The early conservation movement in the United States was also successful due to the hard work of John Muir. Muir was a former carriage worker who was nearly blinded by an accident at work. After almost losing his sight, Muir decided to see "America's natural wonders". Based upon his travels throughout Yosemite and the Sierra Nevada Mountains, Muir wrote a collection of articles for *Century* magazine, entitled "Studies in the Sierra" In 1892, John Muir joined forces with the editor of "Century" Magazine, Robert Underwood Johnson, to establish the Sierra Club, an organization designed to protect America's natural resources and public parks. Early Americans recognized the importance of natural resources and the necessity of wilderness preservation for sustained yield harvesting of natural resources. In essence, the preservation of wilderness and landscapes were recognized as critical for future generations and their continued subsistence in a healthy environment. The foundation of the conservation movement is grounded during this period between 1850 and 1920. Ultimately, historical trends and cultural mind-sets were united, which influenced ideas and policy towards the early history of the conservation movement in the United States.

Early American Conservation Movement

America had its own conservation movement in the 19th century, most often characterized by George Perkins Marsh, author of *Man and Nature*. The expedition into northwest Wyoming in 1871 led by F. V. Hayden and accompanied by photographer William Henry Jackson provided the imagery needed to substantiate rumors about the grandeur of the Yellowstone region, and resulted in the creation of Yellowstone National Park, the world's first, in 1872. Travels by later U.S. President Theodore Roosevelt through the region around Yellowstone provided the impetus for the creation of the Yellowstone Timberland Reserve in 1891. The largest section of the reserve was later renamed Shoshone National Forest, and it is the oldest National Forest in the U.S. But it was not until 1898 when German forester Dr. Carl A. Schenck, on the Biltmore Estate, and Cornell University founded the first two forestry schools, both run by Germans. Bernard Fernow, founder of the forestry schools at Cornell and the University of Toronto, was originally from Prussia (Germany), and he honed his knowledge from Germans who pioneered forestry in India. He introduced Gifford Pinchot, the "father of American forestry", to Brandis and Ribbentrop in Europe. From these men, Pinchot learned the skills and legislative patterns he would later apply to America. Pinchot, in his memoir history Breaking New Ground, credited Brandis especially with helping to form America's conservation laws.

Pinchot wrote that the principles of conservation were:

— *Development*: "the use of the natural resources now existing on this continent for the benefit of the people who live here now. There may be just as much waste in neglecting the development and use of certain natural resources as there is in their destruction. ... The development of our natural resources and the fullest use of them for the present generation is the first duty of this generation."

— *Conservation*: "...the prevention of waste in all other directions is a simple matter of good business. The first duty of the human race is to control the earth it lives upon."

— *Protection of the public interests*: "The natural resources must be developed and preserved for the benefit of the many, and not merely for the profit of a few.

In 1891, Congress passed the Forest Reserve Act, which allowed the President of the United States to set aside forest lands on public domain A decade after the Forest Reserve Act, presidents Harrison, Cleveland, and McKinley had transferred approximately 50,000,000 acres (200,000 km^2) into the forest reserve system. However, President Theodore Roosevelt is credited with the institutionalization of the conservation movement in the United States.

For President Roosevelt, the conservation movement was not about the preservation of nature simply for nature itself. After his experiences traveling as an enthusiastic, zealous hunter, Roosevelt became convinced of "the need for measures to protect the game species from further destruction and eventual extinction" President Roosevelt recognized the necessity of carefully managing America's natural resources. According to Roosevelt, "We are prone to speak of the resources of this country as inexhaustible; this is not so". Nonetheless, Roosevelt believed that conservation of America's natural resources was for the successful management and continued sustain yield harvesting of these resources in the future for the benefit and enjoyment of the American people. Roosevelt took several major steps to further his conservation goals.In 1902, Roosevelt signed the National Reclamation Act, which allowed for the management and settlement of a large tract of barren land Then, in 1905, President Roosevelt helped to create the United States Forest Service and then appointed respected forester, Gifford Pinchot, as the first head of the agency. By the end of his presidency, Theodore Roosevelt, in partnership with Gifford Pinchot, had successfully increased the number of national parks as well as added area to existing forest reserves.

Despite these advancements, the American conservation movement did have difficulties. In the early 1900s the conservation movement in America was split into two main groups: conservationists, like Pinchot and Roosevelt, who were utilitarian foresters and natural rights advocates who wanted to protect forests "for the greater good for the greatest length", and preservationists, such as John Muir, the founder of the Sierra Club. Important differences separated conservationists like Roosevelt and Pinchot from preservationists like Muir. As a preservationist, Muir envisioned the maintenance of pristine natural environments where any development was banned Whereas conservationists wanted regulated use of forest lands for both public activities and commercial endeavors, preservationists wanted forest to be preserved for natural beauty, scientific study and recreation. The differences continue to the modern era, with sustainable harvest and multiple-use the major focus of the U.S. Forest Service and recreation emphasized by the National Park Service.

Index

D

E

F

G

H

I

J

K

L

M

N

O

P

R

S

T

U

W

Y

❑❑❑